Game Theory for
Wireless Engineers

Game Theory for Wireless Engineers
Allen B. MacKenzie and Luiz A. DaSilva

978-3-031-00544-2   paper   Mackenzie/DaSilva
978-3-031-01672-1   ebook   Mackenzie/DaSilva

DOI   10.1007/978-3-031-01672-1

A Publication in the Springer series
SYNTHESIS LECTURES ON COMMUNICATIONS
Lecture #1

First Edition
10 9 8 7 6 5 4 3 2 1

Game Theory for Wireless Engineers

Allen B. MacKenzie and Luiz A. DaSilva
Virginia Polytechnic Institute and State University

SYNTHESIS LECTURES ON COMMUNICATIONS #1

ABSTRACT

The application of mathematical analysis to wireless networks has met with limited success, due to the complexity of mobility and traffic models, coupled with the dynamic topology and the unpredictability of link quality that characterize such networks. The ability to model individual, independent decision makers whose actions potentially affect all other decision makers makes game theory particularly attractive to analyze the performance of ad hoc networks.

Game theory is a field of applied mathematics that describes and analyzes interactive decision situations. It consists of a set of analytical tools that predict the outcome of complex interactions among rational entities, where rationality demands a strict adherence to a strategy based on perceived or measured results. In the early to mid-1990's, game theory was applied to networking problems including flow control, congestion control, routing and pricing of Internet services. More recently, there has been growing interest in adopting game-theoretic methods to model today's leading communications and networking issues, including power control and resource sharing in wireless and peer-to-peer networks.

This work presents fundamental results in game theory and their application to wireless communications and netwokring. We discuss normal-form, repeated, and Markov games with examples selected from the literature. We also describe ways in which learning can be modeled in game theory, with direct applications to the emerging field of cognitive radio. Finally, we discuss challenges and limitations in the application of game theory to the analysis of wireless systems. We do not assume familiarity with game theory. We introduce major game theoretic models and discuss applications of game theory including medium access, routing, energy-efficient protocols, and others. We seek to provide the reader with a foundational understanding of the current research on game theory applied to wireless communications and networking.

KEYWORDS

Game theory, wireless networks, communications theory, wireless communications, distributed protocols, optimization.

To Our Mothers
Eliette Maria N. Pereira da Silva (1942–2004)
Sara June Bryan MacKenzie (1941–)

Contents

Preface

A great many books have been written about game theory. Most of them adopt either an economic perspective or a mathematical perspective. In the past several years, though, the application of game theory to problems in communications and networking has become fashionable and, occasionally, productive. Specifically, game theoretic models have been developed to better understand congestion control, routing, power control, topology control, trust management, and other issues in wired and wireless communication systems. The promise of game theory as a tool to analyze wireless networks is clear: By modeling interdependent decision makers, game theory allows us to model scenarios in which there is no centralized entity with a full picture of network conditions. The challenges in applying game theory as a tool, though, are sometimes less clear. Game theoretic models, like other mathematical models, can abstract away important assumptions and mask critical unanswered questions. It is our hope that this book will illuminate both the promise of game theory as a tool for analyzing networks and the potential pitfalls and difficulties likely to be encountered when game theory is applied by practicing engineers and researchers.

We have not attempted to cover either game theory or its applications to wireless communications and networks in encyclopedic depth. We have severely restricted our exposition to those topics that we feel are necessary to give the reader a grounding in the fundamentals of game theory and its applications to wireless systems, while providing some pointers to other sources that can provide more detail.

In Chapter 1, we introduce the topic of game theory and its applications, without delving into mathematical detail. In Chapter 2, we present the fundamentals of decision and utility theory. These topics are often omitted from introductory game theory texts on the grounds that students of economics will study them in other courses. We do not expect our readers to be students of economics, though, and a failure to grasp this foundational material can lead to errors in the application of game theory. In Chapter 3, we introduce strategic form games. In Chapter 4, we extend this framework to include repeated games, and in Chapter 5 we discuss potential games and other results regarding convergence to equilibrium. In Chapter 6, we address future research directions. Throughout, we attempt to integrate applications of the material to problems in wireless communications and networking.

We thank all of those who have played a part in the preparation of this text. Our series editor, Bill Tranter, inspired us to attempt this project in this new format and also provided detailed comments on the entire book. Our publisher, Joel Claypool, was patient with our missed deadlines and responsive to our questions and concerns. The reviewers were generous and kind with their comments; we hope that we have addressed their suggestions at least in part. We are also grateful to our colleagues and students who contributed to this book in many ways.

We also thank those who ignited our interest in game theory and networking and have continued to support our endeavors therein: Steve Wicker and Larry Blume (for ABM) and Victor Frost and David Petr (for LAD).

This material is based in part on the work supported by the Office of Naval Research and the National Science Foundation under Grant Nos. N000140310629 and CNS-0448131, respectively. Any opinions, findings, and conclusions or recommendations expressed in this material are our own and do not necessarily reflect the views of the Office of Naval Research or the National Science Foundation. We are grateful to these sponsors for their support.

Allen Brantley MacKenzie
Blacksburg, Virginia

Luiz Antonio DaSilva
Arlington, Virginia

CHAPTER 1

Introduction to Game Theory

The purpose of this book is to introduce wireless engineers to game theory. In recent years, there has been a great deal of interest in applying game theory to problems in wireless communications and networking. We strive to give researchers and practitioners the tools that they need to understand and participate in this work, to provide some guidance on the proper and improper uses of game theory, and to examine potential future directions and applications of game theory.

The literature on game theory is vast, and we will certainly not do it justice in this short book. Instead, we have chosen a set of topics from game theory that we feel are particularly relevant to the modeling of wireless networks, and we have covered those topics in as much generality as we feel appropriate for those applications. (Probably the most comprehensive treatment of game theory is [1], although even it does not cover more recent developments.) Nor have we written a comprehensive survey of prior efforts to apply game theory to problems in networking or communications. Our purpose here is strictly tutorial: We present some topics from game theory and show how they might be used to address problems in wireless communications and networking.

This chapter will introduce the general ideas of game theory, briefly describe its history, argue that it is an important subject for wireless researchers to study, and discuss how to avoid common errors made in applying game theory. It will also provide a brief introduction to some example applications of the theory, and will provide a brief summary of the topics to be addressed in the later chapters.

1.1   WHAT IS GAME THEORY?

> Game theory is a bag of analytical tools designed to help us understand the phenomena that we observe when decision-makers interact.
>
> —Martin Osborne and Ariel Rubinstein [2]

Game theory provides a mathematical basis for the analysis of interactive decision-making processes. It provides tools for predicting what might (and possibly what should) happen when agents with conflicting interests interact. It is not a single monolithic technique, but a collection of modeling tools that aid in the understanding of interactive decision problems.

A game is made up of three basic components: a set of players, a set of actions, and a set of preferences. The players are the decision makers in the modeled scenario. In a wireless system, the players are most often the nodes of the network. The actions are the alternatives available to each player. In dynamic or extensive form games, the set of actions might change over time. In a wireless system, actions may include the choice of a modulation scheme, coding rate, protocol, flow control parameter, transmit power level, or any other factor that is under the control of the node. When each player chooses an action, the resulting "action profile" determines the outcome of the game.

Finally, a preference relationship for each player represents that player's evaluation of all possible outcomes. In many cases, the preference relationship is represented by a utility function, which assigns a number to each possible outcome, with higher utilities representing more desirable outcomes. In the wireless scenario, a player might prefer outcomes that yield higher signal-to-noise ratios, lower bit error rates, more robust network connectivity, and lower power expenditure, although in many practical situations these goals will be in conflict. Appropriately modeling these preference relationships is one of the most challenging aspects of the application of game theory.

Examples of games in the real world include conventional games of strategy such as chess and poker, strategic negotiations such as those for the purchase of an automobile, house, or business, as well as daily group decision-making processes such as deciding where to go to lunch with colleagues or what movie to see with your family or friends.

A clear distinction should be drawn, however, between a game, which must involve multiple decision makers, and an optimization problem, which involves only a single decision maker. A game model is usually appropriate only in scenarios where we can reasonably expect the decisions of each agent to impact the outcomes relevant to other agents.

A single shopper buying groceries at a grocery store is performing an optimization by attempting to maximize satisfaction with the items purchased within his or her budget. Although the grocery store is a decision maker in this scenario by setting the prices, it is most likely a passive decision maker, unlikely to be swayed by the actions of a single consumer. On the other hand, a grocery store may be engaged in pricing games with other stores in the area, and the store will respond to the aggregate decisions of consumers. In this scenario, a game model may be appropriate.

Similarly, a single car navigating a roadway in an attempt to reach a destination as quickly as possible performs an optimization. When the roadway is shared, though, the drivers are engaged in a game in which each attempts to reach her destination as quickly as possible without getting in an accident or receiving a traffic ticket.

For a more conventional type of game, consider the game represented by the following table. This is a two-player game. Player 1 chooses the row, and player 2 chooses the column. The values in each cell give the utilities of each player; these utilities represent the players' preferences. The first number listed is the utility of player 1; the second is the utility of player 2. This particular game has a parameter c, which varies between 0 and 1.

	T	W
T	$(-c, -c)$	$(1-c, 0)$
W	$(1-c, 0)$	$(0, 0)$

One of the goals of game theory is to predict what will happen when a game is played. The most common prediction of what will happen is called the "Nash Equilibrium." A Nash equilibrium is an action profile at which no player has any incentive for unilateral deviation.

The game shown above has two Nash equilibria. The Nash equilibria are the action profiles (T,W) and (W,T). Consider the action profile (T,W). In this case, player 1 plays the action T and receives a utility of $1-c$ while player 2 plays the action W and receives a utility of 0. Player 1 has no incentive to deviate, because changing her action to "W" would decrease her utility from $1-c$ (a positive number) to 0. Player 2 has no incentive to deviate, because changing her action to "T" would decrease her utility from 0 to $-c$.

On the other hand, consider a non–Nash Equilibrium action profile, (T,T). From this action profile, player 1 could increase her utility by unilaterally changing her action to W. Such a unilateral deviation would change the action profile to (W,T), thereby increasing player 1's utility from $-c$ to $1-c$.

Chapter 3 will provide a formal mathematical definition of Nash Equilibrium and touch on such philosophical questions as why and how Nash Equilibria might be expected to arise when a game is played.

1.2   WHERE DID GAME THEORY COME FROM?

Game theory's roots are extremely old. The earliest investigations of probability, which often focused on games of chance, might be considered game theory, although they may fail our test of involving multiple decision makers. Even without resorting to that, though, in 1985 it was recognized that the Talmud (0–500 AD) anticipates some of the results of modern game theory [3]. In modern times, work by Cournot, Edgeworth, Zermelo, and Borel in the nineteenth and early twentieth centuries lays the groundwork for the analysis of strategic games. (For a time line of the history of game theory, which provided basic information for this entire section, see [4].)

Nonetheless, modern game theory usually traces its roots to the seminal *Theory of Games and Economic Behavior*, by John von Neumann and Oskar Morgenstern, published in 1944. Von Neumann and Morgenstern gave an axiomatic development of utility theory, which now dominates economic thought and which we will explore in the next chapter, and also introduced the formal notion of a cooperative game. Like much early work in game theory, though, their work on noncooperative game theory (which now dominates the game theory literature) focused on relatively simple two-person zero-sum games.

In papers between 1950 and 1953, John Nash contributed to the development of both noncooperative and cooperative game theory [5–7]. His most important contribution, though, was probably his existence proof of an equilibrium in noncooperative games, the Nash Equilibrium [5]; we will examine this proof in some detail in Chapter 3. For his contributions to game theory, Nash shared in the 1994 Bank of Sweden Prize in Economic Sciences in Memory of Alfred Nobel.

This early work was significantly aided by work completed at the Rand Corporation, which was formed after the World War II to connect military planning with research. With this charter, the Rand Corporation sought to develop theories and tools for decision making under uncertainty, leading to important contributions not only in game theory but also in linear and dynamic programming and mathematical modeling and simulation. One of the early applications of this theory was the study of the decision making of nations during the cold war [8].

Since many of the early game theorists were economists, it is not surprising, perhaps, that game theory's greatest success has been in revolutionizing the field of economics. Ideas developed in the context of game theory now pervade almost every aspect of economic thought: utility theory, strategic use of information, cooperative and team games, the problem of coordination between independent players, auction theory, and implementation of incentive mechanisms.

Game theory has also made significant contributions to other disciplines, including political science, sociology, biology, and military strategy. In the next section, we will discuss why we believe that game theory has the potential to make significant contributions to the study of wireless communications and networking.

1.3   WHY IS GAME THEORY RELEVANT TO WIRELESS COMMUNICATIONS AND NETWORKING?

Ad hoc networks have occupied a preeminent place in the wireless communications and networking literature for the last several years. An ad hoc network is a self-configuring, multihop network in which there is no central authority. Thus, every aspect of the configuration and operation of an ad hoc network must be completely distributed. Furthermore, nodes in an ad hoc network are often severely energy and power constrained. In emerging wireless networks, such as sensor networks, mesh networks, and pervasive computing systems, many of these same features—decentralized operation, self configuration, and power/energy awareness—are often desirable.

Game theory, as we have seen, is the study of the interaction of autonomous agents. At this point, it should be clear how game theory may be of help when analyzing modern wireless networks. In a modern wireless network, each node running a distributed protocol must make its own decisions (possibly relying on information from other nodes). These decisions may be constrained by the rules or algorithms of a protocol, but ultimately each node will have some leeway in setting parameters or changing the mode of operation. These nodes, then, are autonomous agents, making decisions about transmit power, packet forwarding, backoff time, and so on.

In making these decisions, what does the node seek to optimize? In some cases, nodes may seek the "greater good" of the network as a whole. In other cases, nodes may behave selfishly, looking out for only their own user's interests. In a final case, nodes may behave maliciously, seeking to ruin network performance for other users. In the second and third cases, the application of game theory may be straightforward, as game theory traditionally analyzes situations in which player objectives are in conflict. In the first case, node objectives may be aligned (as all players seek the "greater good" of the network), but game theory may still offer useful insights. Even when nodes have shared objectives, they will each have a unique perspective on the current network state, leading to possible conflicts regarding the best course of action.

1.4   HOW CAN I USE GAME THEORY PROPERLY?

There are many potential pitfalls on the road to the application of game theory to wireless communications and networking. One of these pitfalls has already been mentioned—mistaking a simple optimization problem for a game. As we have already discussed, a problem is only a game if there are multiple agents involved in making decisions. In some cases, it may be that artificially creating multiple agents to play a game is worthwhile; in many cases, though, one would be better served by using an optimization algorithm or technique. So, if it is not clear who the agents are, one should carefully consider whether or not the use of game theory is warranted.

A second common mistake in applications of game theory is confusion between the theories of cooperative and noncooperative games. In this book, we focus on noncooperative games, which have been primary in the study of game theory for the last 25 years or so. Noncooperative game theory assumes agents with distinct interests that interact through a predefined mechanism; it encompasses concepts such as normal and extensive form games, incomplete information, Nash equilibrium, and forward and backward induction. There is also a substantial body of literature on cooperative games, in which players form coalitions and negotiate via an unspecified mechanism to allocate resources. Cooperative game theory encompasses concepts such as the Nash bargaining solution and the Shapely value. Although there are certainly profitable applications of both theories and deep connections between the two, the amateur game theorist would be well advised to mix the two theories with extreme caution. Mixing the two theories can easily lead to results that are incomprehensible, at best, and utter nonsense, at worst.

Perhaps the most common mistake among legitimate applications of game theory, though, is a failure to clearly define the game and the setting in which it is played. Who are the players of the game? What actions are available to the players? What are the players' objectives? Does an equilibrium exist for the given game? Is it unique? Is there a dynamic process by which players update their strategies? If so, what is it and is it guaranteed to converge?

Finally, there are two general "philosophies" on applying game theory, and it is extremely important to consider which philosophy you are using. In many cases, researchers who have not clearly thought through their reasoning will jump back and forth between the two philosophies. This is bound to create holes in their arguments, as the philosophies are mutually exclusive.

The first philosophy might be called a direct application of game theory. In this philosophy, the users' actual preferences are said to be modeled by the players' utility functions. What makes the application of this philosophy difficult is that it is very difficult to extract preferences from users. Furthermore, nailing down a single explicit utility function that represents the preferences of all users is not possible, since different users will have different preferences. It is unlikely that even a parameterized utility function can represent the preferences of every user except in very simple (perhaps contrived) situations. Even if one could postulate the existence of such a function, it would take considerable effort to prove that such a formulation was valid for the majority of users in the system.

Thus, if one wants to claim that the utility functions in a game represent actual users' preferences, then one must operate with a bare minimum collection of assumptions about the users' preferences. Typical assumptions might include an assumption that users prefer higher signal-to-interference-and-noise ratios (SINRs) to lower SINRs and that users prefer to spend less power rather than more to obtain a given SINR. This approach can produce beautiful results. It is, however, a very challenging task to execute well.

The second philosophy might be called an "engineering" application of game theory. This philosophy assumes that the engineer is capable of programming the devices in the system to behave *as if* they are maximizing a chosen utility function. Since the engineer *chooses* the utility function, it can be whatever the engineer desires. The burden in this philosophy lies in explaining *why* a particular utility function has been chosen and *why* game theory is being employed at all. Though perhaps easier than demonstrating that a particular utility function represents the preferences of all users, this may still be a difficult challenge to meet.

Note that there is a stark contrast between these two philosophies. Either you are claiming that your utility functions represent actual user preferences—in which case you had better provide some evidence or make the assumptions about those functions very light—or you are not—in which case you had better provide some other compelling reason for using game theory.

The point of this section is not to discourage the use of game theory. Obviously, the authors think that game theory is a valuable tool for approaching a number of different problems; as discussed before, game theory has revolutionized economics. The point, though, is that game theory, like any tool, must be used carefully and on the right kind of problems. To quote a familiar cliché, "If the only tool you have is a hammer, then every problem looks like a nail." We must be vigilant not to fall into that trap.

1.5   INTRODUCTION TO EXAMPLES

In this work, we will discuss a variety of communications and networking examples to illustrate the concepts that we introduce. These examples should provide a sense of how game theory has been used by researchers and will hopefully inspire you to consider some of your own game theory applications. The examples should also demonstrate (on a small scale) each of the two philosophies discussed above.

This section briefly introduces three examples to start your thinking along game theoretic lines.

1.5.1   Power Control

The problem of uplink power control in a code division multiple access (CDMA) cellular system has been considered by a number of authors [9–14]. For simplicity, we will discuss a single-cell model, although multi-cell models (and, more recently, even ad hoc models) are also discussed in the literature.

The players in a cellular power control game are the cellular telephones in the cell. This is often assumed to be a fixed, known number of players. Each player's action is the power level that she chooses. In many cases, this is limited to be a power level between 0 and $p_{\max}$, although some games consider power levels between 0 and ∞.

The question of payoff functions, as usual, is the most difficult question; we will consider it further in the next chapter. For now, though, consider the following properties, which make this problem appropriate for a game model.

- The player's payoff is a function of her own transmit power level and her SINR. The player's SINR is a function of her own transmit power and the transmit powers of the other players in the cell.

- When a player increases her power level, this will increase her own SINR, but will decrease the SINRs of all other players.

- For a fixed SINR, the players prefer lower power levels to higher ones. That is, players wish to conserve power and extend their battery life when possible.

- For a fixed power level, players prefer higher SINR to lower SINR. That is, players want the best possible channel conditions for a given expenditure of power.

Game theoretic analysis of this situation has yielded several conclusions. First, taken as a simple game, the power control situation described here usually leads to extremely inefficient outcomes. What typically occurs is that each player increases her power to increase her SINR. This power increase, though, decreases the SINRs of other users, who then increase their power levels to compensate. By the time an equilibrium is reached, all users are blasting away at much higher power levels than necessary—they could all reduce their powers and everyone would be better off. This could perhaps be viewed as a negative result for game theory, but we believe that it provides some insight into the reality of power control: left to their own devices, selfish users will behave inefficiently in the power control game.

Second, though, game theory provides some insight into ways to obtain better outcomes in the power control scenario. For instance, it is possible to add external incentives against power increases, such as charging users based on transmit power. Another option is to model the scenario as a repeated game, where users can punish each other for using too much power. We will look at some of these possibilities in a future chapter.

What makes this problem interesting to model as a game? There are multiple nodes making individual power control decisions that impact the performance of every node in the cell. Thus, we have an interactive decision problem of the type that is well modeled by game theory.

1.5.2   Routing

Another problem that is well modeled by game theory is the problem of routing in a network. Some examples of work in this area include [15–18]. For simplicity, we assume that a policy of source routing is followed, where the complete route for each packet is determined by the packet source.

In this problem, the players in the game could be viewed as the source nodes in the network, but it is slightly more convenient to view a player as a source/destination pair. (In reality, the decision making will be carried out by the source, but making the source/ destination pair the player allows for the existence of multiple flows from a single source.) The action set available to each player is the set of all possible paths from the source to the destination. (In many formulations of the routing game, nodes can actually choose multiple routes and decide how much of their flow to send on each route. We ignore this complication for now.)

Preferences in this game can take several forms, but we will assume that the preferences are determined by the end-to-end delay for a packet to traverse the chosen route. A short delay is preferred to a longer delay.

If a network contains only a single source and destination pair, or if the available routes were completely disjoint, this problem would represent a simple optimization problem. In the realistic scenario, though, many of the links in the network may potentially be shared by multiple routes. Presumably, the more flows use a given link, the higher the delay will be on that link. It is this dependency that is the source of the game theoretic routing problem.

One of the most interesting results to come from consideration of this type of problem is the Braess paradox. Suppose that a given network has reached equilibrium. One might assume that if you add additional links to the network, then performance will improve, at least, on average. Such an assumption would be incorrect, though, as one can readily generate examples where the addition of new links to the network actually degrades network performance for all users. This phenomena, where more resources lead to worse performance, is known as the Braess paradox [19, 20].

1.5.3   Trust Management

In networks where individual nodes are expected to perform services on behalf of others, the question of trust and reputation management becomes important. This question arises in peer-to-peer (P2P) networks, grid networks, and wireless ad hoc networks, among others. Perhaps most relevant to the topic of this book are ad hoc networks, where each node serves as source/destination for traffic as well as a router to forward packets for its neighbors. What are the incentives for nodes to cooperate in such environments, particularly when there may be natural disincentives such as increased energy consumption?

One possible approach is for each node to assign a reputation value to all other nodes with which it comes into contact in the network. As a node's perceived reputation decreases, its neighbors may refuse to perform services for it, leading to its gradual exclusion from the network. Examples of this type of approach include [21–26].

This problem has many of the characteristics that would lead to a natural game theoretic formulation. Nodes decide independently the extent of their cooperation of the network, trying to balance their reputation (too little cooperation might lead to becoming a pariah in the network) and resource considerations (too much cooperation may lead to fast depletion of their battery). There is also clear impact of a node's decision on other nodes' performance.

Game theory has been applied to the problem of node cooperation, with a repeated game formulation that also analyzed the impact of rogue nodes in the network [27]. The effectiveness of trust and reputation mechanisms to ensure node cooperation was also studied using a game theoretic model in [28] and [29].

1.6   NOTATION

In general, we adopt the established game-theoretic convention of referring to generic players in a game with female pronouns. In cases where it is clear that our players are inanimate objects (such as network nodes), though, we may use gender-neutral pronouns.

We refer to sets with bold capital letters. For instance, a player's action set might be denoted $\mathbf{A_i}$. A generic element of a set is often represented by the same letter as the set, but in lowercase, e.g., $a_i \in \mathbf{A_i}$. Cartesian products of sets are often represented by the same uppercase bold letter, but without the subscript, e.g., $\mathbf{A} = \times_{i \in \mathbf{N}} \mathbf{A_i}$. Bold lowercase letters represent vectors, e.g., since $\mathbf{A}$ is a Cartesian product of sets, an element of $\mathbf{A}$ will be a vector, $\mathbf{a} \in \mathbf{A}$.

Bold capital letters are also used to represent matrices. It should be clear from the context whether a given symbol represents a matrix or a set.

We also follow the widespread game theory convention that uses $-i$ to indicate that a given symbol refers to all players except for player i. So, if $\mathbf{A}$ is defined as above, then $\mathbf{A_{-i}} = \times_{j \in \mathbf{N}, j \neq i} \mathbf{A_j}$ and $\mathbf{a_{-i}} \in \mathbf{A_{-i}}$. Furthermore, if $\mathbf{a} \in \mathbf{A}$, then $\mathbf{a_{-i}} \in \mathbf{A_{-i}}$ indicates the same vector as $\mathbf{a}$, but with the ith element removed. Hence, if $b_i \in \mathbf{A_i}$, then we might write $(\mathbf{a_{-i}}, b_i) \in \mathbf{A}$ to indicate a vector that differs from $\mathbf{a}$ only in that the ith element has been replaced with b_i.

Finally, we adopt the usual mathematical convention of sometimes writing $x \Rightarrow y$ to mean "if x then y." We also use the notation $x \Leftrightarrow y$ to mean "x if and only if y."

1.7   OUTLINE OF REMAINING CHAPTERS

Chapter 2 takes a step back from game theory to explore the subject of decision theory. Game theory is in many ways the multiplayer analogue of decision theory—decision theory considers the choices of a single player, while game theory considers the choices of multiple players. In this chapter we will start with the principle of preferences and discuss what is required to move from preferences to utility functions (the representation of preferences commonly used in game theory).

Chapter 3 then introduces the basics of game theory in more formal terms than we have employed so far. It provides a mathematical definition of a game and of equilibrium and proves the existence of equilibria.

Chapter 4 deals with games that have stages: repeated games and Markov games. These types of games are very important for wireless applications. Repeated games are important because many decisions in a wireless network (such as power level, route, and trust decisions as discussed above) are made repeatedly. Markov games are important because of their explicit representation of the "state" of a game. Since this game state can correspond to the state of a link or network, this type of game is very valuable for wireless applications.

Chapter 5 addresses convergence to equilibrium in a setting very similar to that of the repeated game. It attempts to provide an answer to a question raised in Chapter 3 regarding how Nash Equilibria may come about when games are played. Specifically, we look at the best and better reply dynamics and their convergence to equilibrium in a specific class of games.

Finally, Chapter 6 addresses future directions for applications of game theory to wireless networks.

CHAPTER 2

Decision Making and Utility Theory

Before we dive headlong into game theory, it is worthwhile to spend one chapter studying decision theory. In many ways, game theory is nothing more than multiagent decision theory. So, it is nearly impossible to really understand the former without a grasp of the latter.

Thus, in this chapter we will study classical decision theory, leading up to the work of von Neumann and Morgenstern. Our context will be the preferences of users of wireless communications devices. These might be physical layer preferences for things like low transmit power or a high signal-to-interference-and-noise ratio (SINR). Or they might be network layer preferences for robust routes, or application layer preferences for long battery life and high-quality voice and video.

2.1   PREFERENCE RELATIONSHIPS

Let $\mathbf{X}$ be any set, called the set of outcomes or alternatives, and let $\succeq$ be a binary relation on $\mathbf{X}$. The binary relationship in which we are interested is called a preference relationship, and $x \succeq y$ is read "x is weakly preferred to y" where $x, y \in \mathbf{X}$.

A binary relation $\succeq$ on $\mathbf{X}$ is said to be *complete* if for all $x, y \in \mathbf{X}$ either $x \succeq y$ or $y \succeq x$ (or both). (Note that throughout this text, "or" does not preclude the possibility that both items occur.) A binary relation is said to be *transitive* if $x \succeq y$ and $y \succeq z$ implies that $x \succeq z$.

Definition 1.   *The binary relation $\succeq$ is a* (weak) preference relation *if it is complete and transitive.*

Even at this early stage, we have already begun to sow the seeds of controversy. On the surface, completeness and transitivity seem like reasonable properties for a preference relationship to possess. And, from the perspective of a computer or a programmed agent, they are probably reasonable. It is not at all clear, though, that human preferences satisfy these properties.

First, completeness requires that we express preferences between any two items in the set of outcomes even if they are totally different. Let x be a connection with a throughput of 100 kbps and a delay of 10 ms. Let y be a connection with a throughput of 100 Mbps and a delay of 1 s. Completeness says that either $x \succeq y$ or $y \succeq x$. Now, this probably makes sense if we have a specific application in mind. For example, $x \succeq y$ for real-time voice traffic, but $y \succeq x$ for streaming video. It immediately makes it clear, though, that our preferences depend on context. We can no longer even postulate that we could generate a preference order that would apply for every application.

For a more extreme (nonwireless) example, completeness requires that consumers express preferences between Apple^{TM} computers and oranges, despite the fact that they satisfy completely different needs.

Second, transitivity ultimately requires that we make very fine distinctions. Let us introduce some more notation to clarify this example. Given a weak preference relation $\succeq$, we define strong preference and indifference. We say that $x \succ y$ if $x \succeq y$ and $y \not\succeq x$; this is strong preference. We say that $x \sim y$ if $x \succeq y$ and $y \succeq x$; this is indifference.

Now suppose that you prefer one teaspoon of sugar in your coffee. That is to say that x, a cup of coffee with one teaspoon of sugar, is strictly preferred to y, a cup of coffee with no sugar, or $x \succ y$. Presumably, however, you are indifferent between a cup of coffee containing no sugar, y, and a cup containing one grain of sugar z_1, $y \sim z_1$. And you probably can not tell the difference between one grain (z_1) and two grains (z_2), either $z_1 \sim z_2$. In fact, if we let z_n be a cup of coffee containing n grains of sugar, then one suspects that $z_2 \sim z_3$, $z_3 \sim z_4$, ..., $z_n \sim z_{n+1}$. But the transitivity of $\succeq$ implies that $\sim$ is also transitive. By transitivity, then $y \sim z_1 \sim \ldots \sim z_{19999} \sim x$,[1] or $y \sim x$, but this contradicts our earlier statement that $x \succ y$.

In wireless terms, if our choice set contains connections with various throughputs, and if we prefer higher throughputs to lower throughputs, then we must prefer a 100.000001 Mbps connection to a 100.000000 Mbps connection, even though such a distinction seems foolish!

What do preferences have to do with wireless systems? Well, we presume that users have a preference relationship over outcomes. At the application layer, users prefer high-quality video to low quality video. They prefer a responsive browsing experience to one that requires much waiting. These application layer preferences might be translated into network layer preferences for good connectivity to the network and robust routes to reach important network destinations. Finally, these preferences might be translated into physical layer preferences for high SINR and low bit error rate (BER). Furthermore, depending on the application being run, the user's preferences may change. When transferring an important file, the user may require an extremely low BER, on the order of 10^{-12}, regardless of the connection speed. When streaming a video, though, the user may tolerate a higher BER in order to maintain a given bit rate.

2.2   EXISTENCE OF ORDINAL UTILITY REPRESENTATIONS

Preference relationships are a messy business. If the comparison set is large or infinite, then fully defining the preference relationship could require a lot of information in the form of a very long list of preferences. After all, we know from completeness that for every x, $y \in \mathbf{X}$, we must

---

[1]A teaspoon contains about 20,000 grains of sugar, according to the British Sugar Web site.

have either $x \succeq y$ or $y \succeq x$ (or both). If we could come up with a numerical representation of $\succeq$, then this would greatly simplify the process of specifying the preference relationship.

Definition 2. *A preference relationship $\succeq$ is said to be* represented by a utility function, $u : \mathbf{X} \to \mathbb{R}$, *when*

$$x \succeq y \Leftrightarrow u(x) \geq u(y).$$

Under what conditions can we find such a function u? That is the question to be addressed in this section.

2.2.1   Finite X

It turns out that when $\mathbf{X}$ is a finite set, our life is quite easy.

Theorem 1. *The binary relation $\succeq$ on the finite set $\mathbf{X}$ is a preference relation if and only if there exists a utility function u that represents $\succeq$.*

Proof. We will begin by assuming the existence of a utility function and proving that the binary relation must be a preference relation. Let $\mathbf{X}$ be a finite set and let $\succeq$ be a binary relation that is represented by the function u.

For any $x, y \in \mathbf{X}$, we have two real numbers $u(x)$ and $u(y)$. Since the real numbers form a totally ordered set, we must have either $u(x) \geq u(y)$ or $u(y) \geq u(x)$. Thus, since the function u represents the binary relation $\succeq$, we must have either $x \succeq y$ or $y \succeq x$. Therefore, $\succeq$ is complete.

If $x \succeq y$ and $y \succeq z$, then we must have $u(x) \geq u(y) \geq u(z)$. Hence, $u(x) \geq u(z)$, so that $x \succeq z$. Therefore, $x \succeq z$, and so the binary relation $\succeq$ is transitive. Since $\succeq$ is complete and transitive, it is a preference relation.

Now, we will prove that if there exists a preference order $\succeq$ on a finite set $\mathbf{X}$, then we can define a utility function u that represents $\succeq$.

Suppose that $\succeq$ is a preference relation on the finite set $\mathbf{X} = \{x_1, x_2, \ldots, x_N\}$. For any $x \in \mathbf{X}$, define the less preferred set $\mathbf{L}(x) = \{y \in \mathbf{X} : x \succeq y\}$. Now, let

$$u(x) = |\mathbf{L}(x)|.$$

That is, $u(x)$ is the number of items in the set $\mathbf{L}(x)$.

Now, if $x \succeq y$, then by transitivity, if $z \in \mathbf{L}(y)$, then $z \in \mathbf{L}(x)$. So, $\mathbf{L}(y) \subseteq \mathbf{L}(x)$. Thus $u(x) \geq u(y)$. That is, $x \succeq y$ implies $u(x) \geq u(y)$.

On the other hand, suppose that $x \not\succeq y$. Then by completeness, $y \succeq x$, hence by the argument in the previous paragraph, $\mathbf{L}(x) \subseteq \mathbf{L}(y)$. But $y \in \mathbf{L}(y)$ and $y \notin \mathbf{L}(x)$, thus $\mathbf{L}(x)$ is a strict subset of $\mathbf{L}(y)$ and so $u(y) > u(x)$. Or, in other words, $u(x) \not\geq u(y)$.

Therefore $x \succeq y$ if and only if $u(x) \geq u(y)$. $\square$

Thus, every preference relationship on a finite set $\mathbf{X}$ can be represented with a utility function u.

2.2.2 Countable X

The situation is nearly the same if the choice set $\mathbf{X}$ is infinite, but countable, although the proof is slightly more involved. (A set is said to be countably infinite if there is a one-to-one correspondence between the elements of the set and the integers. The rational numbers are countable, but the real numbers are not. For a proof of this fact, and further discussions of countable infinities versus uncountable infinities, see any text on real analysis.)

Theorem 2. *The binary relation $\succeq$ on the countable set $\mathbf{X}$ is a preference relation if and only if there exists a utility function u that represents $\succeq$.*

Proof. The first part of the proof that the existence of a utility function representing $\succeq$ implies that $\succeq$ is a preference relationship goes through exactly as before. That portion of the previous proof did not make use of the fact that $\mathbf{X}$ was finite.

So, all that remains is to show that if $\succeq$ is a preference relation on a countable set $\mathbf{X}$, then there exists a utility function u that represents $\succeq$. Our proof will follow the same lines as before, with a slight modification to ensure that u remains finite.

Let $\mathbf{X} = \{x_1, x_2, x_3, \ldots\}$. (Note that we can list the elements of $\mathbf{X}$ like this only because $\mathbf{X}$ is countable. It is impossible to array an uncountable set in such a list!) Define the function $v : \mathbf{X} \to \mathbb{R}$ as $v(x_n) = 2^{-n}$. Define the less preferred set $\mathbf{L}(x)$ as before, $\mathbf{L}(x) = \{y \in \mathbf{X} : x \succeq y\}$. Now, define the function u as

$$u(x) = \sum_{y \in \mathbf{L}(x)} v(y).$$

Note that the function u is well-defined because the sum on the right-hand side must always converge (since $\sum_{x \in \mathbf{X}} v(x) = 1$).

Now, we must show that u represents $\succeq$. We proceed as before. If $x \succeq y$, then by transitivity $\mathbf{L}(y) \subseteq \mathbf{L}(x)$. Hence we will have $u(x) \geq u(y)$. If $x \not\succeq y$, then by completeness $y \succeq x$, so that $\mathbf{L}(x) \subseteq \mathbf{L}(y)$. But $y \in \mathbf{L}(y)$ and $y \notin \mathbf{L}(x)$ so that $\mathbf{L}(x)$ is a strict subset of $\mathbf{L}(y)$ and so $u(y) > u(x)$ or $u(x) \not\geq u(y)$. Hence $x \succeq y$ if and only if $u(x) \geq u(y)$, and so u represents $\succeq$. $\qquad\square$

2.2.3 Uncountable X

Finally, we deal with the difficult case of an uncountable choice set $\mathbf{X}$. The importance of this case is debatable. On the one hand, many of the parameters of the channels and network links that we are discussing are, in fact, continuous in nature. Throughput and delay, for instance, are real-valued parameters. For the mathematical purist, it is obvious that one must deal with the technical issue of the existence of utility representations of preference relations on uncountable sets. On the other hand, though, in almost all implementations of communication systems, either the parameter can take only on a restricted number of different values (for instance, only certain signaling and code rates are supported) or we can measure only the parameter to a certain level of accuracy (e.g., delay).

The proof in the previous section will not work if X is uncountably infinite. Why? Because our utility construction in the last section required that we be able to list the elements of X, and we can not construct such a list if X is uncountable.

In fact, we can offer a counter-example to demonstrate that the theorem in the previous section cannot possibly hold if X is uncountable. Let $X = [0, 1] \times [0, 1]$ and define the *lexicographic* preference relation as follows:

$$(x_1, x_2) \succeq (y_1, y_2) \quad \text{if} \quad x_1 > y_1 \quad \text{or} \quad (x_1 = y_1 \quad \text{and} \quad x_2 \geq y_2).$$

Proving that this binary relation is a preference order and that it cannot be represented by a utility function is left as an exercise for the reader. Even if you can not prove that the relation cannot be represented by a utility function, you should at least be able to convince yourself that there is no obvious way to construct such an function.

So, when can we construct a utility function that represents a preference order on an uncountable set X? The answer is technical, requires that we invoke a "trick" from real analysis, and requires a definition.

Definition 3. *Given a binary relation $\succeq$ on X, a subset $A \subseteq X$ is* order dense *in X with respect to $\succeq$ if for all $x \succ y$, there exists $a \in A$ such that $x \succeq a \succeq y$.*

If you are not already familiar with real analysis, this notion of an "order dense" set may seem absurd. In many cases, though, A might be a smaller (i.e., countable rather than uncountable), simpler set than X, and it is often possible to prove that a property of the small set A must also apply to the large set X. The most classic example of the application of order dense subsets is the real numbers. The rational numbers are order dense in the reals with respect to $\geq$. That is, for any two real numbers such that $x > y$, you can find a rational number q, which lies between the two reals: $x \geq q \geq y$.

Theorem 3. *Given a binary relation $\succeq$ on a set X, $\succeq$ has a utility function representation if and only if $\succeq$ is a preference relation and there is a countable set A which is order dense in X with respect to $\succeq$.*

We omit the formal proof. To construct a utility function, though, we can use the same function as in the previous proof, except that we redefine the less preferred set to contain only elements in A. That is, define $L_A(x) = \{y \in A : x \succeq a\}$, and then define the utility function as before, replacing $L(x)$ with $L_A(x)$.

Unfortunately, this theorem is only mildly useful, as it is often difficult to identify an appropriate countable set A and to prove that it is order dense with respect to $\succeq$. As a result, we state the following definition and slightly more useful result.

Definition 4. *The binary relation $\succeq$ on $\mathbf{X}$ is continuous if for all sequences $\{x_n\}$ from $\mathbf{X}$ such that $x_n \to x$: (1) $x_n \succeq y$ for all n implies that $x \succeq y$, and (2) $y \succeq x_n$ for all n implies that $y \succeq x$.*

You should be able to convince yourself that many "real-world" preferences—especially those encountered in wireless networks—will be continuous. An exception to this would be preferences that exhibit a "threshold" of acceptability. For instance, a preference relation where the only requirement is that the throughput be at least 10 Mbps and the user is indifferent to everything other than whether or not this threshold is exceeded.

Note that the definition of continuity requires that we be able to define limits in $\mathbf{X}$. Formally, we require that $\mathbf{X}$ be a separable metric space. If you do not know what that means, do not worry about it. For the purposes of this work, we are not interested in uncountable sets $\mathbf{X}$, which are not separable metric spaces.

Theorem 4. *The binary relation $\succeq$ on $\mathbf{X}$ is a continuous preference relation if and only if there exists a continuous function $u : \mathbf{X} \to \mathbb{R}$ such that $x \succeq y \Leftrightarrow u(x) \geq u(y)$.*

Note that this theorem actually buys us two things. First, it tells us that if $\succeq$ is a preference relation and it is continuous, then a utility representation exists. Second, though, it tells us that this utility function will itself be a continuous function.

While the requirement that preferences be continuous seems reasonable in many instances, from a mathematical point of view it is quite a strong requirement, just as continuity is a strong requirement for functions.

2.2.4 Uniqueness of Utility Functions

Are the utility functions that we have constructed in the above theorems unique? The answer is no. Consider the following lemma.

Lemma 1. *Let u be a utility function that represents the preference relation $\succeq$, and let $f : \mathbb{R} \to \mathbb{R}$ be any strictly increasing function. Then the composite function $f \circ u$ is also a utility representation of $\succeq$.*

Proof. The proof of this lemma is trivial.

Suppose that $x \succeq y$. Since u is a utility representation of $\succeq$, then $u(x) \geq u(y)$. Therefore, $f(u(x)) \geq f(u(y))$, because f is a strictly increasing function.

Suppose that $x \not\succeq y$. Then $u(x) \not\geq u(y)$, or $u(x) < u(y)$. Hence, $f(u(x)) < f(u(y))$, or $f(u(x)) \not\geq f(u(y))$.

Thus $x \succeq y$ if and only if $f(u(x)) \geq f(u(y))$.   □

In fact, we can do even better. Consider the following theorem.

Theorem 5. *Given a preference relation $\succeq$ on a set $\mathbf{X}$ and two utility functions u and v that both represent $\succeq$, there exists a function $f : \mathbb{R} \to \mathbb{R}$ such that $v(x) = f(u(x))$ for all $x \in \mathbf{X}$ and f is strictly increasing on the image of $\mathbf{X}$ under u.*

The results of this section on uniqueness are often summarized by saying that the utility function u generated by our representation theorems is "unique up to an increasing transformation."

2.3 PREFERENCES OVER LOTTERIES

In many cases, users will be expected to have preferences not over certain outcomes but over "lotteries." For instance, one available action may provide an guaranteed SINR of 12 dB while another might provide an SINR of 15 dB with probability 0.9 or an SINR of 5 dB with probability 0.1. In choosing between these two available actions, the user must be able to express a preference between these two outcomes.

Thus, we first need a representation for uncertainty. That is, the objects of choice, $x \in \mathbf{X}$, are no longer simple outcomes, but now represent uncertain prospects. The usual way to mathematically represent uncertainty is with probability. So, we will now let $\mathbf{Z}$ be the set of outcomes, and we will let $\mathbf{X}$, the set of choice objects, be a set of probability distributions on $\mathbf{Z}$. For the purposes of this section, we will assume that a probability distribution is given and is objective, in the sense that we know exactly how to quantify the probabilities involved. The next section will briefly mention some alternatives.

In this setting, we need to be careful with regard to the number of outcomes in $\mathbf{Z}$, as in the last section, and with the probability distributions that we choose to allow. If $\mathbf{Z}$ is finite, then we might allow $\mathbf{X}$ to be the set of all probability distributions on $\mathbf{Z}$. If $\mathbf{Z}$ is uncountably infinite, though, then for technical reasons we must be more careful in our definition of $\mathbf{X}$. In every case, though, we assume that $\mathbf{X}$ is a convex set.

We now want to know, as in the last section, whether or not a utility representation exists for such a preference relation over lotteries. We are not looking for just any utility representation, though. If we were, we could just apply the theorems that we have already stated. Instead, we would like to know whether or not so-called expected utility representations exist. Namely, we want to know if there exists a function defined on $\mathbf{Z}$, the set of prizes, $u : \mathbf{Z} \to \mathbb{R}$, such that

$$p \succeq q \Leftrightarrow E_p[u(z)] \geq E_q[u(z)]$$

where E_p means the expected value with respect to the probability distribution $p \in \mathbf{X}$.

2.3.1   The von Neumann–Morgenstern Axioms

Under what conditions does an expected utility representation for preferences exist? The three von Neumann–Morgenstern axioms are central to answering this question.

Axiom 1. *The binary relation $\succeq$ on $\mathbf{X}$ is a preference relation.*

The first axiom relates back to the previous section, where we saw that for finite and countable $\mathbf{X}$, this axiom was necessary and sufficient for the existence of a utility representation. It is not surprising, then, that it is necessary for the existence of an expected utility representation.

Axiom 2. *For all $p, q, r \in \mathbf{X}$ and $a \in [0, 1]$, $p \succeq q$ if and only if $ap + (1 - a)r \succeq aq + (1 - a)r$.*

Note that this axiom introduces the notion of taking convex combinations of items in $\mathbf{X}$. Recall that we earlier assumed that $\mathbf{X}$ was a convex set, which makes it possible to construct these convex combinations. We might interpret a convex combination as a compound lottery: $ap + (1 - a)r$ means that with probability a we will receive lottery p and with probability $1 - a$ we will receive lottery r.

The second axiom is known as the independence axiom. Suppose that we flip a coin which comes up heads with probability a and tails with probability $1 - a$. Then the only difference between $ap + (1 - a)r$ and $aq + (1 - a)r$ is which lottery we get when the coin comes up heads—when the coin comes up tails, we get the lottery r. Hence, if $p \succeq q$, we should prefer the first mixture to the second. This axiom is also sometimes known as independence of irrelevant alternatives—our relative ranking of the lotteries p and q, should be independent of the lottery r.

The independence axiom, though, is sometimes challenged because of the fact that human decision makers often violate it. Consider the following restatement of the famous Allais Paradox [30]. Suppose that you are running a streaming video application. Depending on the quality of your connection, there are three possibilities: high-quality video, low-quality video, and no video.

First, you are asked to choose between the following two options: (A) this connection gives high-quality video with probability 0.49 and no video with probability 0.51 and (B) this connection gives low-quality video with probability 0.98 and no video with probability 0.02. Which would you choose?

Second, you are asked to choose between the following two options: (C) this connection gives high-quality video with probability 0.001 and no video with probability 0.999 and (D) this connection gives low-quality video with probability 0.002 and no video with probability 0.998. Again, think about which you would prefer.

In similar experimental settings with money, most players choose B over A and choose C over D [31]. But this result violates the independence axiom: The choice of B over A in combination with the independence axiom implies that getting low-quality video with certainty is preferable to a lottery in which one gets high-quality video with probability 1/2 and no video with probability 1/2. (The "irrelevant alternative" which we have discarded here is the 0.02 probability of getting no video, which applies in both cases.) But preferring C over D implies just the opposite.

Two things are uncertain, however: First, it seems far from certain that a person's attitude about wireless links and networks will be similar to her attitude about money. (Did you fall for the paradox above, or were your choices consistent with the independence axiom?) Second, in any setting it is unclear that a programmed agent should violate the axiom. After all, the axiom seems to be a reasonable normative rule for decision making. Why should a programmed agent duplicate the frailties of human decision making?

Axiom 3. *For all $p, q, r \in \mathbf{X}$ such that $p \succ q \succ r$, there exist $a, b \in (0, 1)$ such that $ap + (1 - a)r \succ q \succ bp + (1 - b)r$.*

This is known as the Archimedean axiom. It says roughly that there are no "exceptionally" good or "exceptionally" bad outcomes. Specifically, no matter how bad r is, if we strictly prefer p to q, then we will accept a gamble that assigns a large probability to p and a small probability to r over one that gives us q with certainty. Similarly, no matter how good p is, if we strictly prefer q to r, then we will accept q with certainty over a gamble that assigns a small probability to p and a large probability to r.

It might occur to you that this Archimedean axiom seems related to our notion of continuity from the previous section. This is certainly true! In fact, if $\succeq$ is a continuous preference relation, then the Archimedean axiom is implied.

Lemma 2. *If a preference relation $\succeq$ on $\mathbf{X}$ is continuous, then it satisfies the Archimedean axiom.*

Proof. A consequence of continuity is that if $x_n \to x$ and $x \succ y$ then there exists N such that for all $n \geq N$, $x_n \succ y$. Similarly, if $x_n \to x$ and $y \succ x$ then there exists N such that for all $n \geq N$, $y \succ x$.

Let

$$p_n = \frac{n - 1}{n} p + \frac{1}{n} r$$

and let

$$r_n = \frac{1}{n} p + \frac{n - 1}{n} r.$$

Then clearly $p_n \to p$ and $r_n \to r$. Hence, by continuity, there exists N_1 such that for all $n \geq N_1$, $p_n \succ q$ and N_2 such that for all $n \geq N_2$, $q \succ r_n$.

Let $a = (N_1 - 1)/N_1$ and $b = 1/N_2$. Then $p_{N_1} = ap + (1 - a)r \succ q \succ bp + (1 - b)r = r_{N_2}$ and so the Archimedean axiom is satisfied. $\quad\square$

It is worth saying slightly more about what we mean by limits in this case, but we will save that discussion for a little later.

As noted previously, continuity seems like a reasonable requirement, and so the Archimedean axiom seems reasonable, too. But consider the following: You will probably agree that a free 10 Mbps WiFi Internet connection is preferable to a dial-up connection, and that both are far preferable to your sudden, untimely death. But the Archimedean axiom implies that you would accept a small probability of death in order to upgrade from a dial-up connection to a free WiFi connection! This may seem absurd. But suppose that you are sitting in a coffee shop using a dial-up connection, and I tell you that just across the street is an identical coffee shop with free WiFi. Will you walk across the street? Quite possibly you will, even though this increases the probability of your sudden, untimely demise in the case that you are hit by a car while crossing the street. So, perhaps the Archimedean axiom is reasonable after all. More to the point, though, in wireless networks we are rarely forced to consider outcomes involving death and destruction; they are not typically included in the set of outcomes, $\mathbf{Z}$.

2.3.2 Von Neumann–Morgenstern and the Existence of Cardinal Utility Representations

We are now (mostly) prepared to state the most important results on the existence of expected utility representations. For brevity, we will omit most of the proofs in this section. If you are interested, see a standard reference in decision theory. A good place to start would be [32]; readers with an appreciation for mathematical rigor are referred to [33].

2.3.2.1 Finite Z

If $\mathbf{Z}$ is a finite set and $\mathbf{X}$ is the set of all probability distributions on $\mathbf{Z}$, then the three axioms given above are all that we need.

Theorem 6. *If $\mathbf{Z}$ is a finite set and $\mathbf{X}$ is the set of probability distributions on $\mathbf{Z}$, then a binary relation $\succeq$ on $\mathbf{X}$ satisfies axioms 1, 2, and 3 if and only if there exists a function $u : \mathbf{Z} \to \mathbb{R}$ such that*

$$p \succeq q \Leftrightarrow E_p[u(z)] \geq E_q[u(z)].$$

Unfortunately, when $\mathbf{Z}$ is not a finite set, things become more complicated.

2.3.2.2 Simple Probability Distributions

In most of our applications, though, $\mathbf{Z}$ will not be finite. The "prizes" in our lotteries will often be particular channel conditions or network conditions and, as we discussed previously, the set of such conditions is usually infinite.

So, what if $\mathbf{Z}$ is not finite? Unlike in the case of ordinal utility, restricting $\mathbf{Z}$ to be countably infinite does not help much in this case. Instead, we need to restrict the probability distributions that we consider in a particular way.

Definition 5. *A probability distribution is said to be* simple *if it has finite support. In other words, a probability distribution p on $\mathbf{Z}$ is simple if there exists a subset $\mathbf{Z_S} \subseteq \mathbf{Z}$ such that $\mathbf{Z_S}$ is a finite set and $p(A) = 0$ for any set $A \subseteq \mathbf{Z}$ such that $A \cap \mathbf{Z_S} = \emptyset$.*

Unfortunately, this is a quite restrictive set of probability distributions. By definition, it cannot include any continuous distribution, such as an exponential or normal distribution, and it does not even include discrete distributions that have countably infinite support (for instance, the Poisson distribution). But, we will overcome this restriction with continuity, as we did in the case of ordinal utility.

Theorem 7. *For any $\mathbf{Z}$, if $\succeq$ is a binary relation defined on the set $\mathbf{X}$ of simple probability distributions on $\mathbf{Z}$, then $\succeq$ satisfies axioms 1, 2, and 3 if and only if there is a function $u : \mathbf{Z} \to \mathbb{R}$ such that*

$$p \succeq q \Leftrightarrow E_p[u(z)] \geq E_q[u(z)].$$

Thus, we can preserve the theorem from the previous section by restricting ourselves to simple probability distributions. Note that $\mathbf{X}$ is still a convex set, as any convex combination of simple probability distributions will also be a simple probability distribution.

2.3.2.3 Nonsimple Probability Distributions

With additional axioms, it is possible to prove a theorem similar to the previous theorem even when nonsimple probability measures are included. Unfortunately, some of the additional axioms are quite technical. It is much more straightforward to reintroduce the assumption of continuity.

Let us recall the definition of continuity, specified slightly to the case when $\mathbf{X}$ is a set of probability distributions, instead of an arbitrary set.

Definition 6. *The binary relation $\succeq$ on $\mathbf{X}$ is* continuous *if for all sequences $\{p_n\}$ from $\mathbf{X}$ such that $p_n \to p$: (1) $p_n \succeq q$ for all n implies that $p \succeq q$, and (2) $q \succeq p_n$ for all n implies that $q \succeq p$.*

Note that the convergence in this definition is now a convergence of probability distributions as opposed to convergence in an arbitrary separable metric space. If you are mathematically inclined, then you may recall that there are several different notions of convergence of a random sequence. These include sure convergence, almost sure convergence, mean square convergence, convergence in probability, and convergence in distribution. Convergence in distribution is more commonly known as "convergence in the weak topology" or "weak convergence" because it is implied by most other defined notions of probabilistic convergence. It is this "weak convergence" that is of interest to us here.

Definition 7. *A sequence of probability distributions* $\{p_n\}$ *is said to converge to* p *in the weak topology if for every bounded continuous function* $f : \mathbf{Z} \to \mathbb{R}$, $E_{p_n}[f(z)] \to E_p[f(z)]$.

With this definition, we can now define continuity in the weak topology:

Definition 8. *The binary relation* $\succeq$ *on* $\mathbf{X}$ *is continuous in the weak topology if for all sequences* $\{p_n\}$ *from* $\mathbf{X}$ *such that* p_n *converges to* p *in the weak topology: (1)* $p_n \succeq q$ *for all n implies that* $p \succeq q$, *and (2)* $q \succeq p_n$ *for all n implies that* $q \succeq p$.

And this is the critical definition to extending our previous theorem to the space of all (Borel) probability distributions on $\mathbf{Z}$. (If you do not know what the space of all Borel probability distributions is, then suffice to say that it contains every probability distribution that an engineer will ever encounter!)

Theorem 8. *A binary relation* $\succeq$ *on* $\mathbf{X}$, *the space of all Borel probability distributions on* $\mathbf{Z}$, *satisfies axiom 1 and axiom 2 and is continuous in the weak topology if and only if there exists a utility function* $u : \mathbf{Z} \to \mathbb{R}$ *such that*

1. $p \succeq q \Leftrightarrow E_p[u(z)] \geq E_q[u(z)]$,
2. *u is a continuous function, and*
3. *u is a bounded function.*

Note that continuity in the weak topology has replaced the Archimedean axiom in our theorem. This is not surprising, as we have already noted that continuity implies the Archimedean axiom.

Also note that our theorem has given us a utility function u that is not only continuous but also bounded. This can actually be somewhat troublesome in real applications, where the most obvious utility functions are often unbounded. Yet our theorem indicates that an unbounded utility function must imply violation of Axiom 1, Axiom 2, or continuity in the weak topology. In fact, it is the continuity in the weak topology that is violated by unbounded utility functions. The easy way around this problem is to restrict $\mathbf{Z}$ to a compact set. In other words, we might restrict our consideration to some maximum attainable signal-to-noise ratio or data rate.

On a more pragmatic note, though, our interest in this chapter is primarily in knowing when it is possible to represent a preference relation with expected utility, rather than the other way around. What we have seen is that Axioms 1, 2, and 3 are necessary for the existence of an expected utility representation, but that they are not sufficient if we want to consider the space of all Borel probability distributions. On the other hand, Axiom 1, Axiom 2, and continuity in the weak topology are sufficient for the existence of a continuous expected utility representation, but that continuity in the weak topology is not necessary, unless you want to ensure that your utility function is bounded. A slightly stronger result, telling us exactly what is required for a continuous (but not necessarily bounded) expected utility relation would be nice, but we have come as far as we can come without delving significantly deeper into mathematical complexity. The level of understanding that we have obtained should be sufficient for our purposes.

2.3.2.4 Uniqueness

There is one more thing that we can say about the utility functions described in this section: they are unique up to a linear transformation. Let us make this claim formal.

Theorem 9. *If $u : \mathbf{Z} \to \mathbb{R}$ represents the binary relation $\succeq$ on $\mathbf{X}$, a set of probability distributions on $\mathbf{Z}$, in the sense that $p \succeq q \Leftrightarrow E_p[u(z)] \geq E_q[u(z)]$, then $v : \mathbf{Z} \to \mathbb{R}$ is another function that also represents $\succeq$ in the same sense if and only if there exist real numbers $a > 0$ and b such that $v(z) = a\,u(z) + b$ for all $z \in \mathbf{Z}$.*

The proof is left as an exercise to the reader. Proving that the existence of a and b guarantees that v represents the same preferences as u is quite obvious, once you recall that expectation, $E[\cdot]$, is a linear operator. The proof in the other direction is more challenging and makes careful use of the Archimedean axiom.

Utility functions in the previous section were unique up to a strictly increasing transformation. Since these utility functions reflect only the ordering of the outcomes, they are sometimes called ordinal utility functions. In this section, the utility functions are more precisely specified, as they are unique up to a linear (or affine) transformation. These utility functions represent not only the ordering of the outcomes in $\mathbf{Z}$ but also the ordering over probability distributions in $\mathbf{X}$. These utility functions are known as cardinal utility functions.

2.4   OTHER VISIONS OF EXPECTED UTILITY REPRESENTATIONS

There exist visions of expected utility representations beyond those provided by von Neumann and Morgenstern. A significant weakness of von Neumann and Morgenstern is that the probabilities provided must be objective measures of probability. In the real world, many "lotteries" that we encounter require a subjective evaluation of probability: What is the probability that the approaching Corvette driver will run the red light? What is the probability that I will lose my job this year? What is the probability that I will receive a vast inheritance from a long lost uncle? In the context of wireless networks, nodes may not have an objective measure of the probabilities involved. What is the probability that a cell phone call made right now in my current location will be dropped? What is the probability that the wireless network in this coffee shop can support my videoconferencing application during the busiest hour of the day? Nodes may have estimates of these probabilities, but they certainly do not have objective values for them.

The key question for decision theorists is whether or not people behave as if they were maximizing expected utility, even in circumstances where objective probabilities are not available. The two major frameworks addressing this question are associated with the names of Savage [34] and Anscombe-Aumann [35].

Without going into detail regarding the two models or the differences between them, suffice to say that under mostly reasonable, but sometimes debatable, assumptions on preferences, players will behave as if they are maximizing an expected utility, even when objective probabilities are not available. The meaning of this result for us is that programming our wireless nodes to estimate probabilities and then attempt to maximize expected utilities should result in node behavior that mimics rationality (provided that the probability estimates and utility functions are reasonable).

We close this chapter, though, by describing the Ellsberg Paradox, which illustrates a failure of the subjective probability approach [36]. Suppose that I show you a large barrel in which I tell you that there are 300 balls, all identical except for color, exactly 100 of which are red. Further more, I tell you that the remaining 200 balls are each either yellow or blue. I am prepared to draw one ball from the urn. I then offer you the following choices.

First, you are asked to choose between (A) a lottery ticket which will give you $1000 if the ball drawn is yellow or blue and (B) a ticket which will give you $1000 if the ball drawn is red or blue. Next, I ask you to choose between (C) a ticket which will give you $1000 if the ball drawn is red and (D) a ticket which will give you $1000 if the ball drawn is yellow.

The vast majority of surveyed participants (including famous economists!) will choose A over B and will choose C over D. But choosing A over B implies that one thinks that the probability of yellow or blue is greater than the probability of red or blue. Now, the probability of yellow or blue is known to be 2/3, and the probability of red is known to be 1/3. Thus ticket A will win with probability 2/3 and ticket B will win with probability $1/3 + P[\text{blue}]$. So, preferring A over B suggests that you believe that $P[\text{blue}] < P[\text{red}] = 1/3 < P[\text{yellow}]$. But choosing C over D implies that you believe that $P[\text{red}] = 1/3 > P[\text{yellow}]$. Obviously, this is a contradiction.

The contradiction comes from the fact that people have a natural inclination to prefer known prospects (as given by a choice of yellow/blue with known probability of 2/3 or a choice of red with known probability 1/3) to unknown prospects (as with yellow or red/blue). Now, it seems entirely sensible to expect that our wireless agents should also prefer known to unknown prospects. But, as the Ellsburg paradox shows, this is impossible if we program those agents to make decisions solely on the basis of expected utility.

2.5   CONCLUSION

In this chapter, we have examined the basic results of decision theory. This theory is foundational to economics and seeks to answer the question: When can we represent an agent's preferences mathematically? In this chapter, we have tried to focus on this subject insofar as it has relevance to autonomous agents in a wireless network. Although the original theory deals primarily with human decision making, we stress that some concerns are different when considering the decisions of programmed agents.

Now that we know what it means to assume that a player has a utility function, we are prepared to deal formally with game theory itself. In many applications of game theory, researchers are called upon to make certain assumptions about utility functions. With this chapter as a starting point, we encourage you to consider those assumptions critically by thinking about what such an assumption might mean to the underlying preferences.

CHAPTER 3

Strategic Form Games

Now that we have seen how preference relationships can be expressed as utility functions, we are ready to formally set up a game in strategic form. The process is similar to setting up a classical optimization problem. As in traditional optimization, there is a function that we are trying to maximize: the utility function. What is different here is that one network participant's decisions (for instance, a parameter that can be set independently by each node) potentially affects the utility accrued by everyone else in the network.

In this chapter, we give the mathematical definition of a strategic form game, discuss solution concepts such as the iterative deletion of dominated strategies and the Nash Equilibrium, and provide some discussion of when Nash equilibria exist. We introduce the notion of mixed strategies and prove the existence of mixed strategy equilibria in finite games. Finally, we provide examples of how certain networking problems such as flow control and pricing can be formulated as strategic form games.

3.1 DEFINITION OF A STRATEGIC FORM GAME

This section presents the definition of a strategic form game. We use a simple peer-to-peer file sharing example to illustrate the definition.

A game consists of a principal and a finite set of players $\mathbf{N} = \{1, 2, \ldots, N\}$. The principal sets the basic rules of the game, and each of the players $i \in \mathbf{N}$ selects a strategy $s_i \in \mathbf{S_i}$ with the objective of maximizing her utility u_i. While the individual strategies might more generally be represented as a vector, the majority of our examples concern the setting of a single parameter by a network node, and in the interest of simplicity we denote s_i as a scalar. Consider, for instance, a game to represent the pricing of network resources. The network provider is the principal, who sets prices for various levels of network service, and the users (or applications, or flows) are the players, who decide what grade of service to utilize.

In this book we focus on the class of noncooperative games, where each player selects her strategies without coordination with others. The strategy profile $\mathbf{s}$ is the vector containing the strategies of all players: $\mathbf{s} = (s_i)_{i \in \mathbf{N}} = (s_1, s_2, \ldots, s_N)$. It is customary to denote by $\mathbf{s_{-i}}$ the collective strategies of all players except player i. The joint strategy space (or the space of strategy profiles) is defined as the Cartesian product of the individual strategy spaces: $\mathbf{S} = \times_{i \in \mathbf{N}} \mathbf{S_i}$. Similarly, $\mathbf{S_{-i}} = \times_{j \in \mathbf{N}, j \neq i} \mathbf{S_j}$.

Finally, the utility function characterizes each player's sensitivity to everyone's actions. It is therefore a scalar-valued function of the strategy profile: $u_i(\mathbf{s}) : \mathbf{S} \to \mathbb{R}$.

For the sake of concreteness, let us consider a game of resource sharing, such as might arise in peer-to-peer networks. In peer-to-peer networks, there may be free-riders, who take advantage of files made available by others but never contribute to the network. Clearly, if all users decide to free-ride, the system will not work because no one will make any files available to other users. Modeling these networks in a game theoretic sense allows us to better understand the incentive structures needed to support resource sharing. In our simple model, each player i can decide whether to share her resources with others (we denote that strategy as $s_i = 1$) or to refrain from sharing ($s_i = 0$). The joint strategy space is $\mathbf{S} = \{0, 1\}^N$. Let us assume the cost for each user of making her own resources available for sharing to be 1.5 units, and that the user benefits by 1 unit for each other user in the network who decides to make her resources available (the values of cost and benefit are chosen arbitrarily and turn out not to affect the expected outcome of the game as long as those values are positive). For $N = 3$, this game can be represented in strategic form in Table 3.1, showing the utilities of each player for every possible strategy profile (note that the number of players was chosen for ease of visualization, although all of the discussion here applies for any $N > 1$). A reasonable question is which of these joint actions is the most likely outcome of the game; we treat that question in the next sections. Note that the strategy profile that maximizes the aggregate utility, a possible indication of social welfare from the network point of view, is (1,1,1). It is not clear, though, that there are intrinsic incentives for players to arrive at that strategy. As you may expect, we are interested in determining the likely outcome of this type of game.

We now tackle the question of how to determine the likely outcome of the game. First, we apply the concept of iterative deletion of dominated strategies. For some games, this will be a sufficient predictor of the outcome of the game. Next, we consider the most common game

TABLE 3.1: Strategic form Representation of a Game of Resource Sharing with Three Players. The Numbers in Parentheses Represent the Utilities Accrued by Players 1, 2, and 3, Respectively, for Each Possible Strategy Profile

	$s_2 = 0$	$s_2 = 1$		$s_2 = 0$	$s_2 = 1$
$s_1 = 0$	(0,0,0)	(1,−1.5,1)	$s_1 = 0$	(1,1,−1.5)	(2,−0.5,−0.5)
$s_1 = 1$	(−1.5,1,1)	(−0.5,−0.5,2)	$s_1 = 1$	(−0.5,2,−0.5)	(0.5,0.5,0.5)
	$s_3 = 0$			$s_3 = 1$	

theoretic solution concept: the Nash equilibrium. It can be shown that every finite strategic-form game has a mixed-strategy Nash equilibrium.

3.2 DOMINATED STRATEGIES AND ITERATIVE DELETION OF DOMINATED STRATEGIES

This section presents the most basic notion of "rational play"—the notion of dominated strategies and the iterated deletion of dominated strategies.

In some games it is possible to predict a consistent outcome based on decisions a rational player would make. While there is no common technique that is guaranteed to arrive at a solution for a general game (when such a solution exists), some games can be solved by iterated dominance. Iterated dominance systematically rules out strategy profiles that no rational player would choose.

Let us now consider the game in Table 3.2. Player 1 can choose between moving to the left, to the right, or staying in the middle ($S_1 = \{L, M, R\}$), while player 2 can choose between moving to the left and moving to the right ($S_2 = \{L, R\}$). Notice that, regardless of what player 2 does, it is never a good idea for player 1 to select $s_1 = R$: we say that this strategy is (strictly) dominated by the other two strategies in player 1's strategy set. Assuming, as always, that player 1 is rational, we can eliminate that row from our considerations of the likely outcome of this game. Once we do that, we notice that strategy $s_2 = R$ dominates strategy $s_2 = L$, and therefore it is reasonable for player 2 to select the former. Finally, if player 2 selects strategy $s_2 = R$, we expect player 1 to select $s_1 = M$. By iterative deletion of dominated strategies, we predict the outcome of this game to be strategy profile (M,R).

We can now proceed with a more formal definition of dominated strategies and iterative deletion of dominated strategies:

Definition 9. *A pure strategy s_i is strictly dominated for player i if there exists $s_i' \in S_i$ such that $u_i(s_i', s_{-i}) > u_i(s_i, s_{-i}) \forall s_{-i} \in S_{-i}$. Furthermore, we say that s_i is strictly dominated with respect to $A_{-i} \subseteq S_{-i}$ if there exists $s_i' \in S_i$ such that $u_i(s_i', s_{-i}) > u_i(s_i, s_{-i}) \forall s_{-i} \in A_{-i}$.*

TABLE 3.2: Left/Middle/Right Game: An Illustration of Dominated Strategies.

	$s_2 = L$	$s_2 = R$
$s_1 = L$	(1,1)	(0.5,1.5)
$s_1 = M$	(2,0)	(1,0.5)
$s_1 = R$	(0,3)	(0,2)

Using these definitions, we define notation for the undominated strategies with respect to $\mathbf{A_{-i}}$ available to player i:

$$\mathbf{D_i(A_{-i})} = \{s_i \in \mathbf{S_i} | s_i \text{ is not strictly dominated with respect to } \mathbf{A_{-i}}\}.$$

We next define notation for the undominated strategy profiles with respect to a set of strategy profiles $A \subseteq S$:

$$\mathbf{D(A)} = \times_{i \in \mathbf{N}} \mathbf{D_i(A_{-i})}.$$

The term $\mathbf{D(S)}$ represents the set of all strategy profiles in which no player is playing a dominated strategy. Likewise, $\mathbf{D^2(S)} = \mathbf{D(D(S))}$ represents the set of all strategy profiles in which no player is playing a strategy that is dominated with respect to the set of undominated strategy profiles $\mathbf{D(S)}$. The sets $\mathbf{D^3(S)}$, $\mathbf{D^4(S)}$, ... are similarly defined, and it can be shown that ... $\subseteq \mathbf{D^4(S)} \subseteq \mathbf{D^3(S)} \subseteq \mathbf{D^2(S)} \subseteq \mathbf{D(S)}$.

The set $\mathbf{D^\infty(S)} = \lim_{k \to \infty} \mathbf{D^k(S)}$ is well defined and nonempty. This set is known as the set of serially undominated strategy profiles, or, more colloquially, as the set of strategy profiles that survive the iterated deletion of dominated strategies.

This set is the first approximation to a "solution" to the game, as it is clear that if rational, introspective players are playing the game, they should not play a strategy that would be eliminated by the iterated deletion of dominated strategies. Unfortunately, for some very well known games $\mathbf{D^\infty(S)}$ may be equal to S, yielding no predictive power whatsoever! (The paper–scissors–rock game in the next section is an example of such a game.) Hence, we need a stronger predictive notion. We therefore move to the broader concept of Nash equilibria, while noting that every Nash equilibrium (in pure strategies) is a member of the set $\mathbf{D^\infty(S)}$. First, though, we introduce mixed strategies.

3.3 MIXED STRATEGIES

This section defines the concept of mixed strategies.

Thus far, we have been assuming that each player picks a single strategy in her strategy set. However, an alternative is for player i to randomize over her strategy set, adopting what is called a *mixed strategy*. For instance, in the file sharing example, a player could decide to share her files with some probability $0 < p < 1$.

We denote a mixed strategy available to player i as σ_i. We denote by $\sigma_i(s_i)$ the probability that σ_i assigns to s_i. Clearly, $\sum_{s_i \in \mathbf{S_i}} \sigma_i(s_i) = 1$. Of course, a pure strategy s_i is a degenerate case of a mixed strategy σ_i, where $\sigma_i(s_i) = 1$.

The space of player i's mixed strategies is Σ_i. As before, a mixed strategy profile $\sigma = (\sigma_1, \sigma_2, \ldots, \sigma_N)$ and the Cartesian product of the Σ_i forms the mixed strategy space Σ. We note that the expected utility of player i under joint mixed strategy σ is

TABLE 3.3: Strategic form Representation of Paper–Scissors–Rock Game

	$s_2 =$ Paper	$s_2 =$ Rock	$s_2 =$ Scissors
$s_1 =$ Paper	(0,0)	(1,−1)	(−1,1)
$s_1 =$ Rock	(−1,1)	(0,0)	(1,−1)
$s_1 =$ Scissors	(1,−1)	(−1,1)	(0,0)

given by

$$u_i(\sigma) = \sum_{s \in S} \left(\prod_{j=1}^{N} \sigma_j(s_j) \right) u_i(s). \qquad (3.1)$$

Following terminology adopted in the context of random variables, it is convenient to define the *support* of mixed strategy σ_i as the set of pure strategies to which it assigns positive probability: supp $\sigma_i = \{s_i \in \mathbf{S_i} : \sigma_i(s_i) > 0\}$.

There are numerous games where no pure strategy can be justified (more precisely, where there are no equilibria in pure strategies), and where the logical course of action is to randomize over pure strategies. Let us take as an example the well-known paper–scissors–rock game, whose strategic form representation is shown in Table 3.3. Following tradition, a rock will break scissors, scissors cut paper, and paper wraps rock. The logical strategy for this game, time-tested in playgrounds all over the world, is to randomize among the three pure strategies, assigning a probability of $\frac{1}{3}$ to each.

We now tackle the question of how to determine the likely outcome of the game. In some cases, the concept of iterative deletion of dominated strategies will be a sufficient predictor of the outcome of the game. For most games, though, we will want a sharper prediction. Next, then, we consider the most common game-theoretic solution concept: The Nash equilibrium.

3.4   NASH EQUILIBRIUM

This section presents the most well-known equilibrium concept in game theory—the Nash equilibrium. Equilibria are found for the example games from earlier sections. We also discuss the shortcomings of the Nash equilibrium—the most glaring of which is the fact that it is almost impossible to justify *why* players in a real game would necessarily play such an equilibrium.

The Nash equilibrium is a joint strategy where no player can increase her utility by unilaterally deviating. In pure strategies, that means:

Definition 10. *Strategy* $\mathbf{s} \in \mathbf{S}$ *is a* Nash equilibrium *if* $u_i(\mathbf{s}) \geq u_i(\hat{s}_i, \mathbf{s}_{-i}) \ \forall \ \hat{s}_i \in \mathbf{S_i}$, $\forall i \in \mathbf{N}$.

An alternate interpretation of the definition of Nash equilibrium is that it is a mutual best response from each player to other players' strategies. Let us first define the *best-reply correspondence* for player i as a point-to-set mapping that associates each strategy profile $\mathbf{s} \in \mathbf{S}$ with a subset of $\mathbf{S_i}$ according to the following rule: $\mathbf{M_i}(\mathbf{s}) = \{\arg\max_{\hat{s}_i \in \mathbf{S_i}} u_i(\hat{s}_i, \mathbf{s_{-i}})\}$. The best-reply correspondence for the game is then defined as $\mathbf{M}(\mathbf{s}) = \times_{i \in \mathbf{N}} \mathbf{M_i}(\mathbf{s})$.

We can now say that strategy $\mathbf{s}$ is a Nash equilibrium if and only if $\mathbf{s} \in \mathbf{M}(\mathbf{s})$. Note that this definition is equivalent to (and, indeed, a corollary of) Definition 10.

What are the Nash equilibria for our previous three examples? Let us start with the resource sharing example in Table 3.1. Joint strategy (0,1,0) is not an equilibrium because player 2 can improve her payoff by unilaterally deviating, thereby getting a payoff of 0 (greater than -1.5). Systematically analyzing all eight possible joint strategies, we can see that the only one where no player can benefit by unilaterally deviating is (0,0,0). This is the only Nash equilibrium for this game. Note that joint strategy (1,1,1) would yield higher payoff than strategy (0,0,0) for every player; however, it is not an equilibrium, since each individual player would benefit from unilaterally deviating. The Nash equilibrium for this game is clearly inefficient, a direct result of independent decisions without coordination among players. Readers who are familiar with basic concepts in game theory will recognize our example of resource sharing as a version of the famous Prisoner's Dilemma [37].

For the example in Table 3.2, iterative deletion of dominated strategies has yielded the unique Nash equilibrium for that game. As discussed before, the paper–scissors–rock admits no Nash equilibrium in pure strategies, but there is an equilibrium in mixed strategies. We must therefore refine our discussion of Nash equilibrium to account for the possibility that a mixed strategy may be the equilibrium.

3.4.1   Dealing with Mixed Strategies

Let us generalize the previous discussion by taking into consideration mixed strategies. We begin by revisiting the concept of best reply:

Definition 11. *The* best reply correspondence in pure strategies *for player $i \in \mathbf{N}$ is a correspondence* $\mathbf{r_i} : \Sigma \rightrightarrows \mathbf{S_i}$ *defined as* $\mathbf{r_i}(\sigma) = \{\arg\max_{s_i \in \mathbf{S_i}} u_i(s_i, \sigma_{-i})\}$.

This definition describes a player's pure strategy best response(s) to opponents' mixed strategies. However, it is possible that some of a player's best responses would themselves be mixed strategies, leading us to define:

Definition 12. *The* best reply correspondence in mixed strategies *for player $i \in \mathbf{N}$ is a correspondence* $\mathbf{mr_i} : \Sigma \rightrightarrows \Sigma_i$ *defined as* $\mathbf{mr_i}(\sigma) = \{\arg\max_{\sigma_i \in \Sigma_i} u_i(\sigma_i, \sigma_{-i})\}$.

Not surprisingly, these two definitions are related. It can be shown [38] that $\hat{\sigma}_\mathbf{i}$ is a best reply to $\sigma_{-\mathbf{i}}$ if and only if supp $\hat{\sigma}_\mathbf{i} \subset \mathbf{r_i}(\sigma)$.

We can now expand on our previous definition of Nash equilibrium to allow for mixed strategy equilibria:

Definition 13. *A mixed strategy profile $\sigma \in \Sigma$ is a Nash equilibrium if $u_i(\sigma) \geq u_i(s_i, \sigma_{-i}) \; \forall \, i \in$* **N**, *$\forall \, s_i \in$ **S$_\mathbf{i}$***.

Finally, we note that σ is a Nash equilibrium if supp $\sigma_\mathbf{i} \subset r_i(\sigma) \; \forall i \in$ **N**.

3.4.2 Discussion of Nash Equilibrium

The Nash equilibrium is considered a consistent prediction of the outcome of the game in the sense that if all players predict that a Nash equilibrium will occur, then no player has an incentive to choose a different strategy. Furthermore, if players start from a strategy profile that is a Nash equilibrium, there is no reason to believe that any of the players will deviate, and the system will be in equilibrium provided no conditions (set of players, payoffs, etc.) change. But what happens if players start from a nonequilibrium strategy profile? There may still be a process of convergence to the Nash equilibrium which, when reached, is self-sustaining. Establishing convergence properties for some of the games we study may be of crucial importance. (We tackle this issue in Chapter 5.) More troubling is the case where multiple Nash equilibria exist: is one of them a more likely outcome of the game than others? Are we assured to converge to any of them? Furthermore, a Nash equilibrium may be vulnerable to deviations by a coalition of players, even if it is not vulnerable to unilateral deviation by a single player.

Despite its limitations, the Nash equilibrium remains the fundamental concept in game theory. Numerous refinements to this concept have been proposed, and we will discuss some of those in the context of repeated games in the next chapter.

In general, the uniqueness or even existence of a Nash equilibrium is not guaranteed; neither is convergence to an equilibrium when one exists. Sometimes, however, the structure of a game is such that one is able to establish one or more of these desirable properties, as we discuss next.

3.5   EXISTENCE OF NASH EQUILIBRIA

This section discusses some of the key results on the existence of a Nash Equilibrium. In particular, we describe the proof for the existence of Nash equilibria for finite games (finite number of players, each of which with a finite strategy set), and we briefly present an existence result for infinite games with continuous payoffs. The section also provides the reader with a general approach to proving equilibrium existence: fixed point theorems.

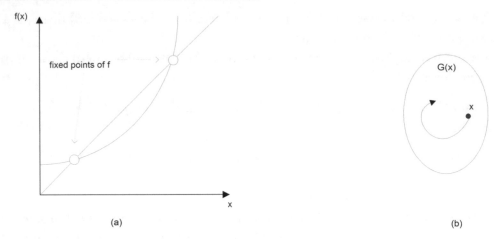

FIGURE 3.1: Illustration of fixed points: (a) for a function; (b) for a correspondence

As fixed point theorems are key to establishing the existence of a Nash equilibrium, we start by reviewing the concept of a fixed point:

Definition 14. *Consider a function with identical domain and range: $f : \mathbf{X} \to \mathbf{X}$. We say that $x \in \mathbf{X}$ is a* fixed point *of function f if $f(x) = x$.*

This definition can be generalized to apply to point-to-set functions (i.e., correspondences):

Definition 15. *Consider a correspondence that maps each point $x \in \mathbf{X}$ to a set $\phi(x) \subset \mathbf{X}$. Denote this correspondence by $\phi : \mathbf{X} \rightrightarrows \mathbf{X}$. We say x is a* fixed point *of ϕ if $x \in \phi(x)$.*

We illustrate fixed points for a function and for a correspondence in Fig. 3.1. How do fixed points relate to Nash equilibria? Consider that player i's best response in mixed strategies, $\mathbf{mr_i}(\sigma)$. Let us define $\mathbf{mr}(\sigma)$ as the Cartesian product of the $\mathbf{mr_i}(\sigma)$. Then, by the definition of the Nash equilibrium as a mutual best response, we see that any fixed point of $\mathbf{mr}$ is a Nash equilibrium. Even more, any Nash equilibrium is a fixed point of $\mathbf{mr}$.

Before we proceed, we need one more definition.

Definition 16. *A correspondence ϕ from a subset $\mathbf{T}$ of Euclidean space to a compact subset $\mathbf{V}$ of Euclidean space is* upper hemicontinuous *at point $x \in \mathbf{T}$ if $x_r \to x$, $y_r \to y$, where $y_r \in \phi(x_r) \, \forall r$, implies $y \in \phi(x)$. The correspondence is* upper hemicontinuous *if it is upper hemicontinuous at every $x \in \mathbf{T}$.*

The well-known proof of the existence of a Nash equilibrium that we present here utilizes Kakutani's fixed point theorem, which can be stated as:

Theorem 10 (Kakutani). *Let* $X \subset \mathbb{R}^m$ *be compact and convex. Let the correspondence* $\phi : X \rightrightarrows X$ *be upper hemicontinuous with nonempty convex values. Then* ϕ *has a fixed point.*

To prove that every finite strategic-form game has a Nash equilibrium, we must therefore prove that Σ is compact and convex and that the best reply correspondence **mr** is upper hemicontinuous with nonempty convex values. Let us do this in several steps.

Lemma 3. Σ *is compact and convex.*

Proof. The Cartesian product of compact, convex sets is compact and convex, so it suffices to show that all Σ_i, $i \in N$, exhibit these properties.

It should be clear that Σ_i is closed and bounded, and therefore compact. It is also quite simple to show its convexity.

We can think of σ_i as a vector associating probabilities σ_i^j, $j = 1, 2, \ldots, |S_i|$ to each pure strategy $s_i \in S_i$. Note that $|X|$ denotes the cardinality of set X. Any $|S_i|$-dimensional vector with the following properties belongs to Σ_i:

1. $\sigma_i^j \geq 0$ for all $j \in \{1, 2, \ldots, |S_i|\}$;
2. $\sum_{j=1}^{|S_i|} \sigma_i^j = 1$.

Now take mixed strategies $\alpha_i, \beta_i \in \Sigma_i$ and some $0 \leq \lambda \leq 1$ and form:

$$\lambda \alpha_i + (1 - \lambda)\beta_i = (\lambda \alpha_i^1 + (1 - \lambda)\beta_i^1, \ldots, \lambda \alpha_i^{|S_i|} + (1 - \lambda)\beta_i^{|S_i|})$$

It is clear that $\lambda \alpha_i^j + (1 - \lambda)\beta_i^j \geq 0 \ \forall j$. Also,

$$\sum_{j=1}^{|S_i|} \lambda \alpha_i^j + (1 - \lambda)\beta_i^j = \lambda \sum_{j=1}^{|S_i|} \alpha_i^j + (1 - \lambda) \sum_{j=1}^{|S_i|} \beta_i^j = \lambda + (1 - \lambda) = 1$$

Thus, $\lambda \alpha_i + (1 - \lambda)\beta_i \in \Sigma_i$, making Σ_i convex. $\square$

Next we must show that **mr** is convex: we accomplish this by showing mr_i to be convex, making **mr**, the Cartesian product, also convex.

Lemma 4. *Let* $\mathbf{mr}(\sigma) = \times_{i \in N} \mathbf{mr_i}(\sigma)$. *The correspondence* $\mathbf{mr}(\sigma)$ *is nonempty and convex.*

Proof. When S_i is finite and nonempty for all $i \in N$, the best reply mapping $mr_i \neq \emptyset$ and therefore $\mathbf{mr} \neq \emptyset$.

Consider, for player i, two mixed strategies α_i, $\beta_i \in \mathbf{mr_i}(\sigma)$, and take $0 < \lambda < 1$. To show that $\mathbf{mr_i}$ is convex, we must show that $\lambda\alpha_i + (1-\lambda)\beta_i \in \mathbf{mr_i}(\sigma)$. Notice that

$$
u_i(\lambda\alpha_i + (1-\lambda)\beta_i, \sigma_{-i}) = \sum_{s \in S} \prod_{j=1}^{N} \sigma_j(s_j) u_i(s)
$$

$$
= \sum_{s \in S} (\lambda\alpha_i(s_i) + (1-\lambda)\beta_i(s_i)) \prod_{j=1, j \neq i}^{N} \sigma_j(s_j) u_i(s)
$$

$$
= \lambda \sum_{s \in S} \alpha_i(s_i) \prod_{j=1, j \neq i}^{N} \sigma_j(s_j) u_i(s)
$$

$$
+ (1-\lambda) \sum_{s \in S} \beta_i(s_i) \prod_{j=1, j \neq i}^{N} \sigma_j(s_j) u_i(s)
$$

$$
= \lambda u_i(\alpha_i, \sigma_{-i}) + (1-\lambda) u_i(\beta_i, \sigma_{-i})
$$

So, if α_i and β_i are both best responses to σ_{-i}, then so is their weighted average. $\square$

Finally, we must show that $\mathbf{mr}(\sigma)$ is upper hemicontinuous. Let us now propose the following lemma; the proof follows that presented in [38].

Lemma 5. *The correspondence $\mathbf{mr}$ is upper hemicontinuous.*

Proof. We will proceed by contradiction. Take sequences of pairs of mixed strategies $(\sigma^k, \hat{\sigma}^k) \to (\sigma, \hat{\sigma})$ such that for all $k = 1, 2, \ldots$, $\hat{\sigma}^k \in \mathbf{mr}(\sigma^k)$ but $\hat{\sigma} \notin \mathbf{mr}(\sigma)$.

Since $\hat{\sigma}$ is not a best response to σ, then for some $i \in \mathbf{N}$, $\exists \bar{\sigma}_i \in \Sigma_i$ such that

$$
u_i(\bar{\sigma}_i, \sigma_{-i}) > u_i(\hat{\sigma}_i, \sigma_{-i}).
$$

Because $\sigma^k \to \sigma$, and thus $\sigma_{-i}^k \to \sigma_{-i}$, we can find k sufficiently large to make $u_i(\hat{\sigma}_i^k, \sigma_{-i}^k)$ arbitrarily close to $u_i(\hat{\sigma}_i, \sigma_{-i})$. So, for large enough k:

$$
u_i(\bar{\sigma}_i, \sigma_{-i}^k) > u_i(\hat{\sigma}_i^k, \sigma_{-i}^k),
$$

which establishes a contradiction, as $\hat{\sigma}_i^k$ is a best response to σ_{-i}^k. Thus, $\mathbf{mr}$ must be upper hemicontinuous. $\square$

The three lemmas above, combined with Kakutani's fixed point theorem, establish the following theorem:

Theorem 11 (Nash). *Every finite game in strategic form has a Nash equilibrium in either mixed or pure strategies.*

The existence of equilibria can also be established for some classes of game with infinite strategy spaces. The procedure is usually analogous to the one followed above, starting from one of several fixed point theorems. An example of such a theorem which can be proved with the Kakutani fixed point theorem already stated is the following, attributed independently to Debreu, Glicksburg, and Fan in [1].

Theorem 12. *Consider a strategic-form game with strategy spaces* S_i *that are nonempty compact convex subsets of an Euclidean space. If the payoff functions* u_i *are continuous in* **s** *and quasi-concave in* s_i, *there exists a Nash equilibrium of the game in pure strategies.*

The proof of this theorem is similar to that of the theorem due to Nash, described above. We construct a best reply correspondence in pure strategies for the game, and we show that the compactness of the action spaces and continuity of the utility functions guarantees that this correspondence is nonempty and upper hemicontinuous. Then, we show that the quasi concavity of the utility functions guarantees that the correspondence is convex valued. Finally, we can apply Kakutani's theorem.

3.6 APPLICATIONS

In this section, we provide examples of some networking problems that can be modeled as strategic form games.

3.6.1 Pricing of Network Resources

In networks offering different levels of quality of service (QoS), both network performance and user satisfaction will be directly influenced by the user's choices as to what level of service to request. Much work has been devoted to designing appropriate pricing structures that will maximize network profitability or some social welfare goal such as the sum of users' payoffs. Since each user's choice of service may be influenced not only by the pricing policy but also by other users' behavior, the problem can naturally be treated under a game-theoretic framework, in which the operating point of the network is predicted by the Nash equilibrium. One such formulation is briefly described here (for more details, the reader should refer to [39]).

Pricing can be treated as a game between a network service provider (the principal) and a finite set of users or traffic flows (the players). Different users or flows may have different QoS requirements, and each user must choose what level of service to request among all service classes supported by the network: this choice is the user's strategy s_i. For instance, in priority-based networks, a strategy may be the priority level a user requests for her traffic; in networks that support delay or data rate guarantees, a strategy may be the minimum bandwidth to which a user requests guaranteed access. The tradeoff, of course, is that the higher the level of service requested the higher the price to be paid by the user. QoS-guaranteed flows are accepted as long as the network is capable of supporting users' requested service. The network service provider architects the Nash equilibria by setting the rules of the game: the pricing structure and the dimensioning of network resources.

A user's payoff is determined by the difference between how much a user values a given QoS level and how much she pays for it. The maximization of this payoff, given all other users' service choices, will determine the optimum strategy to be adopted by each user.

To see that each individual player's strategy is impacted by all others' strategies, we can consider a network where service differentiation is accomplished solely through priority schemes. In this situation, when all users are assigned the highest priority, they will individually experience the same performance as if they had all been assigned the lowest priority. In networks supporting resource reservation, similar interdependencies occur.

The weakest link in this kind of formulation, as is often the case, is in the determination of the utility function. It would be difficult, if not impossible, for a network provider to assess user sensitivity to different levels of performance. Nevertheless, it is possible to obtain some results regarding properties of the Nash equilibrium for the game formulated above, under some general conditions on the utility function. For instance, it is reasonable to assume u_i to be monotonic in each of its variables. We would expect utility to monotonically decrease with QoS parameters such as average delay and packet loss ratio, as well as with cost, and monotonically increase with the amount of bandwidth and buffer space available to users. (We also note that strict monotonicity is not likely, since there may be a point beyond which further increases in QoS or available resources may not yield additional benefit to the user.) Also, u_i might reasonably be assumed to be concave in each of its variables, following a diminishing returns argument. We expect a user's marginal utility to decrease with QoS: the better the quality, the less the user is willing to pay for further improvement.

Results regarding the existence and uniqueness of the Nash equilibrium for the game described above, as well as an analysis of Nash equilibria for specific types of networks supporting QoS differentiation, can be found in [39]. Numerous other authors also apply game theory to the problem of network pricing. Cocchi *et al.* [40] study customer decisions in a two-priority network where a fixed per-byte price is associated with each priority class. They determine that class-sensitive pricing can increase user satisfaction with the cost/benefit trade-offs offered by the network. Lazar *et al.* [41] analyze a noncooperative game in which users reserve capacity with the objective of minimizing some cost function. The authors discuss properties of a Nash equilibrium under a dynamic pricing scheme where the price charged per unit bandwidth depends on the total amount of bandwidth currently reserved by other users.

More recent work has focused on pricing of wireless and ad hoc networks, such as in [42, 43].

3.6.2   Flow Control

Flow control, wherein each user determines the traffic load she will offer to the network in order to satisfy some performance objective, is another network mechanism that has been modeled using game theory.

One of the earliest such models was developed in [44]. In that model, a finite number of users share a network of queues. Each user's strategy is the rate at which she offers traffic to the network at each available service class, constrained by a fixed maximum rate and maximum number of outstanding packets in the network. The performance objective is to select an admissible flow control strategy that maximizes average throughput subject to an upper bound on average delay. The authors were able to determine the existence of an equilibrium for such a system.

CHAPTER 4

Repeated and Markov Games

This chapter considers the concepts of repeated games and Markov games and takes a brief dip into the waters of extensive form games. We illustrate these concepts through examples of voluntary resource sharing in an ad hoc network, power control in cellular systems, and random access to the medium in local area networks.

4.1 REPEATED GAMES

Repeated games are an important tool for understanding concepts of "reputation" and "punishment" in game theory. This section introduces the setting of the repeated game, the strategies available to repeated game players, and the relevant notions of equilibria.

In a repeated game formulation, players participate in repeated interactions within a potentially infinite time horizon. Players must, therefore, consider the effects that their chosen strategy in any round of the game will have on opponents' strategies in subsequent rounds. Each player tries to maximize her expected payoff over multiple rounds.

It is well known that some single-stage games result in Nash equilibria that are suboptimal from the point of view of all players. This is true for the Prisoner's Dilemma as well as for the simple peer-to-peer resource sharing game outlined in the previous chapter. The same games, when played repeatedly, may yield different, and possibly more efficient, equilibria. The key is that each player must now consider possible reactions from their opponents that will impact that player's future payoffs. For instance, selfish behavior may be punished and free riders in the resource sharing game may now have an incentive to make their resources available to others. We will expand on this argument in our extended example.

Note that the assumption that the number of rounds is not known a priori to players is often crucial. Otherwise, players may consider their optimal strategy in the last round of the game and then work their way back from there, often arriving at the same equilibrium as for the single stage game. (To be fair, there do exist formulations of finitely repeated games that lead to a more efficient equilibrium than the stage game.)

Before we go any further, let us discuss the extensive form representation of a game, which will be useful as we explore the refinements of the Nash equilibrium to treat games that are played in multiple stages.

4.1.1   Extensive Form Representation

For most of this book, we will stick to the strategic form representation of games. Sometimes, however, it is convenient to express a game in extensive form.

A game in extensive form is represented as a tree, where each node of the tree represents a decision point for one of the players, and the branches coming out of that node represent possible actions available to that player. This is a particularly convenient way to represent games that involve sequential actions by different players. At the leaves of the tree, we specify payoffs to each player from following that particular path from the root. The extensive form representation can also account for different information sets, which describe how much a player knows when she is asked to select an action.

Note that any game in strategic form can also be represented in extensive form (and vice versa). So, the extensive form representation does not necessarily imply that players' actions are taken sequentially.

Let us take, once again, the simple peer-to-peer resource game from Table 3.1 and express it in extensive form in Fig. 4.1. As before, we do not assume that players' actions are sequential, but rather all three players make their decisions about sharing their files or not independently and simultaneously. This is denoted in the figure as a circle containing all player 2 nodes, and the same for all player 3 notes, representing that at the time of decision player 2 does not know which node it is at (the same for player 3).

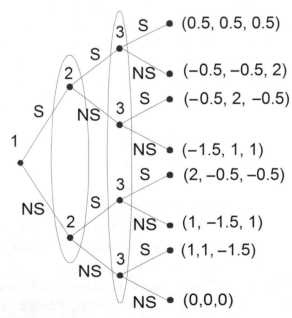

FIGURE 4.1: Extensive form representation of P2P file sharing game

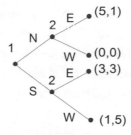

	EE	EW	WE	WW
N	5,1	5,1	0,0	0,0
S	3,3	1,5	3,3	1,5

FIGURE 4.2: Extensive form game example, with actions taken sequentially

What we have seen in the example is a graphical representation of *information sets*. An information set is a collection of decision nodes that are under the control of the same player and which the player is unable to distinguish from one another. In other words, if the player reaches any of those nodes, she will not know which node she has reached.

Let us now construct an example of a sequential game where player's actions *are* taken sequentially: when player 2 selects her actions, she is aware of what action player 1 has taken. Player 1 chooses between two directions (North and South) and, once she made her choice, player 2 then chooses between two directions (East and West). Such a game is shown in both extensive and strategic forms in Fig. 4.2.

The "strategies" available to players in repeated games (or in any extensive form game) are significantly different from the strategies available in strategic form games. A strategy for a player in an extensive form game is a rule that maps every possible information state into an action that the player may take.

In our last example, player 1's strategies can be simply stated as North or South. Player 2's strategies are more complex. One possible strategy is "move East if player 1 moves North but move West if player 1 moves South": this is denoted as strategy EW (the first letter specifies what action player 2 will play if player 1 moves North, and the second letter player 2's action if player 1 moves South). While there are two possible strategies for player 1, there are four strategies available to player 2. The notion of strategy here is strongly related to the information set in which the player operates. What strategies are likely outcomes of this game?

4.1.2   Equilibria in Repeated Games

In addition to the basic notion of Nash equilibria, which can be applied to repeated games virtually unmodified, this section introduces the "subgame perfect equilibrium"—a stronger equilibrium concept that is extremely important in the literature on extensive form games.

Let us go back to the example in Fig. 4.2. Examination of the game in strategic form reveals three Nash equilibria in pure strategies: (N,EE), (N,EW), and (S,WW). The Nash equilibria is most easily determined by examining the game in strategic form to verify that no player can benefit from unilaterally deviating from those strategies. However, are all three equilibria equally likely? In strategy (S,WW), player 2 threatens player 1 with moving West regardless of player 1's action. However, this threat is not credible, as whenever player 1 moves North a rational player 2 should move East: this is sometimes called *cheap talk* from player 2. It is desirable to have a definition of equilibrium for a repeated game that excludes strategies that are based on empty threats.

This leads to a refinement of the concept of Nash equilibrium: the subgame perfect equilibrium. To get there, let us first define a subgame.

Take a node x in an extensive form game Γ^e. Let $\mathbf{F}(x)$ be the set of nodes and branches that follow x, including x. A *subgame* is a subset of the entire game such that the following properties hold:

1. the subgame is rooted in a node x, which is the only node of that information set;

2. the subgame contains all nodes $y \in \mathbf{F}(x)$; and

3. if a node in a particular information set is contained in the subgame, then all nodes in that information set are also contained.

A *proper subgame* of Γ^e is a subgame whose root is not the root of Γ^e. Now, we are ready to define subgame perfect equilibrium:

Definition 17. *A subgame perfect equilibrium $\hat{\sigma}$ of game Γ^e is a Nash equilibrium of Γ^e that is also a Nash equilibrium for every proper subgame of Γ^e.*

Again, let us go back to the example in Fig. 4.2. There are two proper subgames for this game: the subgames are rooted in each of the nodes belonging to player 2. There are three Nash equilibria, but the reader can verify that the only subgame perfect equilibrium is (N,EW).

It should be clear from this example that, while every subgame perfect equilibrium is a Nash equilibrium, the converse is not true.

Let us now introduce some notation for the modeling of repeated games in strategic form.

4.1.3   Repeated Games in Strategic Form

After our brief foray into extensive form games, let us now express a repeated game in strategic form. As before, $\mathbf{N}$ denotes the set of players. We will denote by $\mathbf{A_i}$ the set of actions available to player i in each round of the game. We use the word *action* to emphasize that these refer to decisions the player makes in a given round; this is to differentiate from *strategy*, which refers to rules that map every possible information state the player can be in into an action. Again,

$A = \times_{i \in N} A_i$ and the action profile $\mathbf{a} = (a_i)_{i \in N}$. Since a player can also randomize her actions, it is convenient to define mixed action α_i as a randomization of actions a_i.

The payoff a player receives at a stage of the game, $g_i : A \to \mathbb{R}$, is a function of the action profile. We can collect stage payoffs into $g(\mathbf{a}) = (g_1(\mathbf{a}), \dots, g_N(\mathbf{a}))$. The game $\Gamma = < N, A, \{g_i\}_{i \in N} >$ is referred to as the *stage game*.

We index the actions that players adopt in each round, and the resulting payoffs, by the round number k, $k \in \{0, 1, 2, \dots\}$. In this manner, $\mathbf{a}^k = (a_1^k, a_2^k, \dots, a_N^k)$ is the action profile in the kth stage of the game. We collect players' actions up to stage k into a *history* $\mathbf{h}^k = (\mathbf{a}^0, \mathbf{a}^1, \dots, \mathbf{a}^k)$, $k = 0, 1, 2, \dots$, and $\mathbf{H}^k$ denotes the set of all possible histories up to and including stage k. A pure strategy by player i now assigns an action to be taken in round k to each possible history in the $k - 1$ preceding rounds.

Players strive to maximize their expected payoff over multiple rounds of the game. The resulting payoff is often expressed as a sum of single-round payoffs, discounted by a value $0 \leq \delta < 1$. In this manner, players place more weight on the payoff in the current round than on future payoffs. The average discounted payoff can be expressed as

$$u_i = (1 - \delta) \sum_{k=0}^{\infty} (\delta)^k g_i(\mathbf{a}^k).$$

The $(1 - \delta)$ factor normalizes the average payoff to the same units as the payoff in a stage game. That factor is suppressed if we are interested in the total payoff for the repeated game. The repeated game is sometimes expressed as $\Gamma^r(\delta)$, when one wishes to make it explicit that the game is subject to discount factor δ.

We note that our simple formulation of repeated games presupposes that players' actions in round k are known to all players in round $k + 1$. This is clearly not true in many cases of interest. Let us consider the other extreme: if nothing about opponents' actions in round k is known by players in round $k + 1$, then those actions will not affect players' decisions in subsequent rounds. There is, therefore, no reason to expect the equilibrium to differ from that of the single stage game. There must, however, be some middle ground. Each player may not be aware of her opponent's exact actions in previous rounds but may have available to her some signal that is correlated to those actions. This case is modeled by games of imperfect monitoring, which we briefly discuss in Chapter 6.

It is time for a concrete example to illustrate how playing a game repeatedly may yield payoffs that are more efficient than those supported in a single stage game.

4.1.4   Node Cooperation: A Repeated Game Example

Let us consider in more detail a game of voluntary resource sharing, where all nodes voluntarily perform services directly for one another, helping achieve a network-wide goal. Examples include grid computing, as well as ad hoc, sensor, and peer-to-peer (P2P) networks. Resources being shared may include processing and forwarding capabilities, storage, files, and data aggregation. In many cases of interest, this sharing occurs in a distributed fashion, without a centralized controlling entity.

Cooperation in such environments is often voluntary, with users perceiving some benefit in contribution. For instance, grid computing may rely on users' perception of contributing to a worthy goal by making available their idle CPU cycles for scientific research. However, there are also costs to participating in these environments. For instance, in an ad hoc or sensor network, by forwarding packets for others a node may deplete its own limited energy resources. As their costs outweigh the perceived benefit, users are less likely to volunteer their services, potentially compromising the overall goals of the network. The example discussed in this section is extended in [27].

Consider a repeated game played K times, where K is a geometrically distributed random variable with parameter $0 < p < 1$. We can write $p_k \equiv \text{Prob}[K = k] = p(1 - p)^k$, $k = 0, 1, 2, \ldots$ and therefore $E[K] = \frac{1-p}{p}$. Note that as $p \to 1$ the probability that there will be a next round for the game approaches 0.

We consider homogeneous action spaces for all players $\mathbf{A_i} = \{0, 1\}$, where 1 represents a decision by a user to make her resources available for sharing, while 0 represents the decision to refrain from sharing.

We postulate general properties of the utility function based on general, intuitive assumptions, without attempting to completely characterize such functions. In particular, we consider a user's payoff in round k to be the sum of two components: $g_i(\mathbf{a^k}) = \alpha_i(\mathbf{a^k}) + \beta_i(\mathbf{a^k})$.

Function $\alpha_i(\mathbf{a^k}) = \alpha_i(\sum_{j \in \mathbf{N}, \, j \neq i} a_j^k)$ represents the benefit accrued by a player from her opponents' sharing their resources. We assume $\alpha_i(0) = 0$ and $\alpha_i(\mathbf{a^k}) > 0$ if $\exists \, j \neq i$ such that $a_j \neq 0$, as it is intuitive that a user will accrue positive benefit from others' willingness to freely perform services for her.

On the other hand, there are costs to sharing one's own resources, and those costs are represented by function $\beta_i(\mathbf{a^k}) = \beta_i(a_i^k)$. This function most often takes negative values, representing the cost of participating in the network (such as faster depletion of a node's energy resources); however, occasionally it can be expressed as a positive number, if there exist financial incentives for participation or if the user derives personal satisfaction in doing so. In either case, we assume this part of the utility function to be dependent only on the node's own chosen action. Furthermore, either way, $\beta_i(0) = 0$.

Consider a *grim-trigger strategy* adopted by each node: cooperate as long as all other nodes share their resources; defect if any of the others have deviated in the previous round. The trigger is activated when any one node decides to switch from the desired behavior (sharing resources) and is grim as the nodes do not switch their action back once the punishment (not sharing) is initiated.

Let us consider any round of the game. If a player cooperates, the payoff she should expect from that point forward is

$$[\alpha_i(N-1) + \beta_i(1)][1 + \sum_{k=0}^{\infty} pk(1-p)^k] = \frac{\alpha_i(N-1) + \beta_i(1)}{p}.$$

If, on the other hand, a player deviates, her expected payoff from that round on is simply $\alpha_i(N-1)$. So, the grim trigger strategy is a Nash equilibrium (and, indeed, a subgame perfect equilibrium) if the following inequality holds for all players i:

$$\alpha_i(N-1) > -\frac{\beta_i(1)}{1-p}.$$

The result is simple to interpret. When $\beta_i(1) > 0$, i.e., when the benefit from cooperating in a single round outweighs the cost of doing so, then, not surprisingly, it is always an equilibrium to cooperate. More interestingly, when $\beta_i(1) < 0$, representing a positive cost in cooperating, then a socially optimum equilibrium is also sustainable, subject to the cost/benefit tradeoff represented by the inequality above. Note that such an equilibrium is not sustainable in a single stage game. What is different here is that the threat to defect by opponents makes it rational to cooperate in each stage of the game, provided the player believes that the game will repeat with a high probability.

In this formulation, we chose not to use discounted payoffs. However, the beliefs that a game will be repeated is implicit in the characterization of the number of rounds as a random variable.

This type of formulation also applies to other network environments where nodes are expected to freely provide services for one another. For instance, in multiple-hop ad hoc networks, nodes are expected to forward packets for their neighbors. The willingness to forward packets may be impacted by the threat that refusal to do so will lead others to retaliate. The probability p in our model above would then characterize the likelihood of interacting repeatedly with the same nodes; it could, for instance, depend on node mobility.

This example has natural parallels with the Prisoner's Dilemma, where in a single stage game the only Nash equilibrium is Pareto dominated, but where the repetition of the game can make the Pareto optimal strategy a sustainable equilibrium. However, are those payoffs the only ones achievable in this sort of game? Alas, no. The folk theorems discussed next specify the set of feasible payoffs.

4.1.5   The "Folk Theorems"

This section explains the notion of a "folk theorem"—both as the term is usually used and as it is used with respect to repeated games. We then discuss the most important folk theorem for repeated games. We talk about the "good" aspects of this theorem—there are many equilibria available—and the "bad" aspects—the notion of equilibrium is so watered down as to be almost meaningless!

Folk theorems usually refer to results that are widely believed to be true and passed on orally from generation to generation of theoreticians (hence, the "folk" part) long before they are formally proven. In the context of game theory, the "folk theorems" establish feasible payoffs for repeated games. While these may have originally been "folk theorems" in the conventional sense, the term "folk theorem" is now used in game theory to refer to any theorem establishing feasible payoffs for repeated games. That is to say that calling these theorems "folk theorems" is actually a misnomer, as the theorems have been proven. For this reason, some authors refer to these as general feasibility theorems. The designation of "folk" is colorful but somewhat misleading and not particularly descriptive. Nevertheless, we have chosen to respect the game theory convention by calling these feasibility theorems "folk theorems."

Each folk theorem considers a subclass of games and identifies a set of payoffs that are feasible under some equilibrium strategy profile. As there are many possible subclasses of games and several concepts of equilibrium, there are many folk theorems. We will discuss only two.

Before introducing the folk theorems, we need to go through a series of definitions. We start with the concept of feasible payoffs.

Definition 18.   *The stage game payoff vector* $\mathbf{v} = (v_1, v_2, \ldots, v_N)$ *is* feasible *if it is an element of the convex hull* $\mathbf{V}$ *of pure strategy payoffs for the game:*

$$\mathbf{V} = convex\ hull\ \{\mathbf{u} \mid \exists\ \mathbf{a} \in \mathbf{A}\ such\ that\ g(\mathbf{a}) = \mathbf{u}\}.$$

Recall that the convex hull of a set $\mathbf{S}$ is the smallest convex set that contains $\mathbf{S}$. Or, equivalently, the convex hull of set $\mathbf{S}$ is the intersection of all convex sets that contain $\mathbf{S}$.

Let us define the min–max payoff for player i:

Definition 19.   *The* min–max payoff *for player i is defined as*

$$\underline{v_i} = min_{\alpha_{-i} \in \Delta(\mathbf{A_{-i}})} max_{\alpha_i \in \Delta(\mathbf{A_i})} g_i(\alpha_i, \alpha_{-i})$$

The min–max payoff establishes the best payoff that each player can guarantee for herself, regardless of others' actions. It is therefore sometimes called the *reservation payoff*. Another way to think of the min–max payoff is as the lowest payoff that player i's opponents can hold her to by any choice of action $\mathbf{a_{-i}}$, provided that player i correctly foresees $\mathbf{a_{-i}}$ and plays the best response to it.

It follows that in any Nash equilibrium of the repeated game, player i's payoff is at least $\underline{v_i}$. We need one more definition before we can state a folk theorem:

Definition 20. *The set of* feasible strictly individually rational payoffs *is*

$$\{v \in \mathbf{V} \mid v_i > \underline{v_i} \; \forall \, i \in \mathbf{N}\}$$

A weak inequality in the definition above would define the set of feasible individually rational payoffs, so called because a payoff vector where each player does not receive at least her min–max payoff is clearly not sustainable.

We can now state one of the folk theorems:

Theorem 13. *For every feasible strictly individually rational payoff vector* **v**, $\exists \, \underline{\delta} < 1$ *such that* $\forall \, \delta \in (\underline{\delta}, 1)$ *there is a Nash equilibrium of the game* $\Gamma^r(\delta)$ *with payoffs* **v**.

The intuition behind the folk theorem is that any combination of payoffs such that each player gets at least her min–max payoff is sustainable in a repeated game, provided each player believes the game will be repeated with high probability.

For instance, consider the punishment imposed on a player who deviates is that she will be held to her min–max payoff for all subsequent rounds of the game. In this way, the short-term gain by deviating is offset by the loss of payoff in future rounds. Of course, there may be other, less radical (less grim) strategies that also lead to the feasibility of some of those payoffs.

Let us outline a proof for the theorem. The proof will make use of the following identities, true for $\delta \in [0, 1)$:

$$\sum_{t=k_1}^{k_2} \delta^t = \frac{\delta^{k_1} - \delta^{k_2+1}}{1 - \delta}$$

$$\sum_{t=k_1}^{\infty} \delta^t = \frac{\delta^{k_1}}{1 - \delta}$$

Proof. Take a feasible payoff $\mathbf{v} = (v_1, \ldots, v_N)$, with $v_i > \underline{v_i} \; \forall \, i \in \mathbf{N}$ and choose action profile $\mathbf{a} = (a_1, \ldots, a_N)$ such that $g(\mathbf{a}) = \mathbf{v}$. Consider the following threat: players will play action a_i until someone deviates; once a player deviates, all others will min–max that player for all subsequent rounds. (Notice that due to the definition of Nash equilibrium, it suffices to consider unilateral deviations.)

Can a player benefit from deviating? If player i deviates in round k, she will accrue some payoff $\hat{v}_i = \max_{\hat{a}_i} g_i(\hat{a}_i, \mathbf{a}_{-i})$ in that round and payoff $\underline{v_i}$ thereafter. The average discounted payoff for that strategy is:

$$(1 - \delta)\left[\sum_{t=0}^{k-1} \delta^t v_i + \delta^k \hat{v}_i + \sum_{t=k+1}^{\infty} \delta^t \underline{v}_i\right]$$

$$= (1 - \delta)\left[\frac{(1 - \delta^k)}{1 - \delta} v_i + \delta^k \hat{v}_i + \frac{\delta^{k+1}}{1 - \delta} \underline{v}_i\right]$$

$$= (1 - \delta^k)v_i + \delta^k[(1 - \delta)\hat{v}_i + \delta \underline{v}_i]$$

Of course, the average discounted payoff for *not* deviating is simply v_i. The expression above is less than v_i if $(1 - \delta)\hat{v}_i + \delta \underline{v}_i < v_i$. Finally, since $\underline{v}_i < v_i$, we are guaranteed to find $\delta \in [0, 1)$ that makes the inequality true—we can always choose δ close enough to 1 to make the second term of the sum dominate the first.   □

The attentive reader will notice that we cheated on this proof, as a feasible payoff **v** may require a randomization of actions in the stage game. A more complete proof that also takes care of this can be found in [1].

The result above does not imply the subgame perfection of the strategies employed. A weaker folk theorem addresses the issue of subgame perfect equilibria:

Theorem 14.   *Let there be an equilibrium of the stage game that yields payoffs* $\mathbf{e} = (e_i)_{i \in \mathbf{N}}$. *Then for every* $\mathbf{v} \in \mathbf{V}$ *with* $v_i > e_i$ *for all players i,* $\exists \underline{\delta}$ *such that for all* $\delta \in (\underline{\delta}, 1)$ *there is a subgame-perfect equilibrium of* $\Gamma^r(\delta)$ *with payoffs* $\mathbf{v}$.

The good news from the folk theorems is that we discover that a wide range of payoffs may be sustainable in equilibrium. The bad news is that, if a multiple (infinite!) number of equilibria exist, what hope do we have of making a case that any particular equilibrium is a good prediction of the outcome of the game without coordination among players?

4.2   MARKOV GAMES: GENERALIZING THE REPEATED GAME IDEA

There is a very natural relationship between the notion of a repeated game and that of a Markov game. This section introduces Markov game setting, Markov strategies, and Markov equilibrium. Much of the discussion in this section follows that in [1].

In a *stochastic* or *Markov* game, the history at each stage of the game can be summarized by a *state*, and movement from state to state follows a Markov process. In other words, the state at the next round of the game depends on the current state and the current action profile.

Let us formalize the definition of the game. The game is characterized by state variables $m \in \mathbf{M}$. We must also define a transition probability $q(m^{k+1}|m^k, \mathbf{a}^k)$, denoting the probability that the state at the next round is m^{k+1} conditional on being in state m^k during round k and on the playing of action profile $\mathbf{a^k}$.

Markov strategies, while more complex than the strategies available in a strategic form game, are significantly simpler than the strategies available in repeated games. As in other repeated games, we collect a history of past plays; for Markov games, this history also contains the states traversed at each stage of the game. In other words, the history at stage k is $\mathbf{h^k} = (m^0, \mathbf{a^0}, m^1, \mathbf{a^1}, \ldots, \mathbf{a^{k-1}}, m^k)$. At stage k, each player is aware of the history $\mathbf{h^k}$ before deciding on her action for that stage.

For a generic repeated game, a pure strategy for player i in round k can be denoted as $\mathbf{s_i(h^k)}$; equivalently, a mixed strategy is denoted as $\sigma_i(\mathbf{h^k})$. Recall that the strategies map the entire space of histories into a set of prescribed actions at each round. Markov strategies, however, are simpler, as for each player i and round k, $\sigma_i(\mathbf{h^k}) = \sigma_i(\mathbf{\hat{h}^k})$ if the two histories have the same value of state variable m^k. For this reason, Markov strategies are often denoted $\sigma_i : \mathbf{M} \to \Delta(\mathbf{A_i})$, even though this is a slight abuse of notation.

A Markov perfect equilibrium is a profile of Markov strategies, which yields a Nash equilibrium in every proper subgame. It can be shown [1] that Markov perfect equilibria are guaranteed to exist when the stochastic game has a finite number of states and actions. We also note that the Markov perfect equilibrium concept can be extended to apply to extensive games without an explicit state variable.

Markov chains are used to model a number of communications and networking phenomena, such as channel conditions, slot occupancy in random channel access schemes, queue state in switches, etc. It is natural that Markov games would find particular applications in this field. In particular, the Markov game results imply that if we can summarize the state of a network or a communications link with one or more "state variables," then we lose nothing (in terms of equilibrium existence, anyway) by considering only strategies that consider only these state variables without regard to past history or other information.

4.3   APPLICATIONS

Earlier in the chapter, we outlined an application of repeated games to the issue of selfishness in ad hoc networks (as well as other networks where users are expected to perform services for others). We now briefly describe applications in power control in cellular networks and medium access control in a slotted Aloha system.

4.3.1   Power Control in Cellular Networks

The problem of power control in a cellular system has often been modeled as a game. Let us take a code division multiple access (CDMA) system to illustrate what makes game theory appealing to the treatment of power control.

In a CDMA system, we can model users' utilities as an increasing function of signal to interference and noise ratio (SINR) and a decreasing function of power. By increasing power, the user can also increase her SINR. This would be a local optimization problem, with each user determining her own optimal response to the tradeoff, if increasing power did not have any effect on others. However, CDMA systems are interference limited, and an increase in power by one user may require others to increase their own power to maintain the desired SINR. The elements of a game are all here: clear tradeoffs that can be expressed in a utility function (what the exact utility function should be is another issue) and clear interdependencies among users' decisions.

Let p_i be the power level selected by user i and γ_i the resulting SINR, itself a function of the action profile: $\gamma_i = f(\mathbf{p})$. We can express the utility function as a function of these two factors. One possible utility function is from [45].

$$u_i(\mathbf{p}) = u_i(p_i, \gamma_i) = \frac{R}{p_i}(1 - 2BER(\gamma_i))^L,$$

where R is the rate at which the user transmits, $BER(\gamma_i)$ is the bit error rate achieved given an SINR of γ_i, and L is the packet size in bits.

It has been shown [46] that this problem modeled as a one-stage game has a unique Nash equilibrium. Like the Prisoner's Dilemma and the resource sharing game described earlier, this Nash equilibrium happens to be inefficient.

Now consider the same game played repeatedly. It is now possible to devise strategies wherein a user is punished by others if she selects a selfish action (such as increasing power to a level that significantly impairs others' SINR). If the user's objective is to maximize her utility over many stages of the game, and if the game has an infinite time horizon, the threat of retaliation leads to a Pareto efficient equilibrium.

MacKenzie and Wicker [45] have considered such a game, where cooperating users select a power level that ensures fair operation of the network. That power level can, for instance, be reported by the base station at the beginning of each slot. If a user deviates from that strategy, all others will punish the user by increasing their power levels to the (Pareto-inefficient) values dictated by the Nash equilibrium for the single stage game. Under the assumption that all users are aware of everyone else's power levels at each stage, this strategy will lead to a subgame perfect equilibrium that is more efficient than the Nash equilibrium of the single stage game.

4.3.2   Medium Access Control

In this section, we describe a repeated game of perfect information applied to characterizing the performance of slotted Aloha in the presence of selfish users. This example was first developed by MacKenzie and Wicker [47].

In this model, users compete for access to a common wireless channel. During each slot, the actions available to each user are to transmit or to wait. The channel is characterized by a matrix $\mathbf{R} = [\rho_{nk}]$, with ρ_{nk} defined as the probability that k frames are received during a slot where there have been n transmission attempts. The expected number of successfully received frames in a transmission of size n is therefore $r_n \equiv \sum_{k=0}^{n} k\rho_{nk}$. If we further assume that all users who transmit in a given slot have equal probability of success, then the probability that a given user's transmission is successful is given by $\frac{r_n}{n}$.

This is a repeated game, as for each slot, each user has the option to attempt transmission or to wait. The cost of a transmission is assumed to be $c \in (0, 1)$, with a successful transmission accruing utility of $1 - c$, an unsuccessful transmission utility $-c$ and the decision to wait giving a utility of 0. Payoffs are subject to a per-slot discount factor of $0 \leq \delta < 1$.

A strategy in this game maps from the current number of users in the system (assumed known to all) into a probability of transmission. Players are indistinguishable, so they play the same strategy. MacKenzie and Wicker [47] show that an equilibrium strategy for this game is guaranteed to exist. They also show that there exists a value of c for which the aggregate throughput achieved in this game, with users assumed selfish and making decisions independently, can equal the maximum aggregate throughput for a slotted Aloha system where the decision on who transmits on any given slot is made in a centralized manner.

This characterization of medium access control provides insight into the price of distributed decisions (vs. those controlled by a base station or access point) as well as the impact of different channel models on aggregate throughput expected for random access.

CHAPTER 5

Convergence to Equilibrium: Potential Games

A lingering question from Chapter 3 regarding the Nash equilibrium is why one should ever expect players to play a Nash equilibrium. Perhaps if there exists only one Nash equilibrium, and players are introspective and know the entire game model (including other players' payoffs), an expectation of Nash equilibrium play is warranted. But what about other, more typical, cases? What if there are multiple equilibria? What if players know only their own payoffs and have no awareness of the payoffs of other players? Can we really expect Nash equilibrium play in these cases?

Unfortunately, in general, the answer seems to be "no." The Nash equilibrium may still be valuable to help us understand the stable states of the system, and it may give us some clue as to how to design systems that will converge to more desirable equilibria. But, in arbitrary games there is no known "learning" algorithm that guarantees convergence of play to a Nash equilibrium when players have limited information.

On the other hand, though, there is a significant class of games for which convergence to Nash equilibrium is assured, even if players "learn" via a very simple adaptive process. Furthermore, games from this class seem to appear more often than expected in engineering (and economic) applications. It is this class of games that we will study in this chapter, as we believe that this class has significant promise for further applications and that deeper understanding of this class of games may provide insight into the performance of adaptive player processes in other games as well.

5.1 THE "BEST REPLY" AND "BETTER REPLY" DYNAMICS

Our setting for this discussion is similar to a repeated game. We have a fixed, strategic form game, which is played repeatedly. We denote this strategic form game $\Gamma = (\mathbf{N}, \mathbf{S}, \{u_i(\cdot)\}_{i \in \mathbf{N}})$. In round m of the game, each player selects a strategy $s_i^m \in \mathbf{S_i}$ producing a strategy profile $\mathbf{s^m} \in \mathbf{S}$. (We will use superscripts to denote which round of the game we are discussing.) Unlike those in repeated or Markov games, though, players are not concerned about the future. Instead, players are said to be *myopic*: they care only about their current payoffs, with no concern for the future or the past.

In addition, we assume that players are unable to change strategies in every round. Instead, in each round exactly one player is given an opportunity to change strategies. This player may be chosen randomly, or players may be selected in a round robin fashion. It is worth noting that this requirement is mostly technical; we are simply trying to avoid the situation where multiple players are changing strategies simultaneously. In a real system, where events unfold in continuous time, usually all that is required is that players choose the instants at which they change strategies at random. Then, the probability that two or more players change at the same instant is zero and can be ignored.

Once we know which player will have the opportunity to change strategies in a given round, we need to consider what new strategy the player will choose. It turns out that very simple processes can be very powerful.

Definition 21. *In the* best reply dynamic, *whenever player i has an opportunity to revise her strategy, she will choose*

$$s_i^{m+1} \in \arg\max_{s_i' \in \mathbf{S_i}} u_i(s_i', \mathbf{s_{-i}^m}).$$

In other words, whenever a player has an opportunity to revise her strategy, she will choose a strategy that maximizes her payoff, given the current strategies of the other players.

We now have a dynamic process for updating strategies. In each round, a player is chosen to update her strategy (either in a round robin fashion or at random). When given the opportunity to update, each player will choose a "best response" to the actions of other players. Where will this process lead?

First, we claim that any strategy that survives the best reply dynamic must be a strategy that survives iterated deletion of strictly dominated strategies. The reasoning is simple. Suppose that a strategy is dominated. Then, by definition, it cannot be a best response to any strategy profile. So, as soon as player i has an opportunity to revise her strategy, she will switch to a nondominated strategy. Hence, once everyone has had an opportunity to change strategies, we will be in $\mathbf{D(S)}$, as defined in Chapter 3. Once play is within $\mathbf{D(S)}$, no player will choose an action that is strictly dominated within $\mathbf{D(S)}$. So, by the time everyone has played again, we will be within $\mathbf{D^2(S)}$. This argument can be formalized to show that the play must end up within $\mathbf{D^\infty(S)}$, the set of serially undominated strategy profiles.

This dynamic meets the first definition of rationality, regardless of the specifics of the game. However, all it guarantees us is that the play will end up within the set of serially undominated strategy profiles. The dynamic process may not converge. For example, in the case of a simple game like the paper–scissors–rock game of Chapter 3, if players are chosen to change strategies at random, the play under the best reply dynamic will never converge. Instead, whenever given a chance to change strategies, each player will choose the object that beats the object currently played by her opponent.

However, suppose that the best reply dynamic does converge to a particular strategy profile $\mathbf{s}^* \in \mathbf{S}$. Then $\mathbf{s}^*$ must be a Nash Equilibrium! If it were not a NE, then at least one player would change strategies when given the opportunity.

The best reply dynamic requires that players know the instantaneous payoffs available for each of their strategies, $s_i \in S_i$. An even simpler dynamic is the random better reply dynamic.

Definition 22. *In the* random better reply dynamic, *whenever player i has an opportunity to revise her strategy, she will choose s_i^{m+1} at random from the set $\{s_i' \in \mathbf{S_i} | u_i(s_i', \mathbf{s_{-i}^m}) > u_i(\mathbf{s^m})\}$ unless this set is empty, in which case she will choose $s_i^{m+1} = s_i^m$. In other words, whenever a player has an opportunity to revise her strategy, she will choose a strategy with a higher payoff than her current payoff.*

This strategy can be implemented by randomly sampling another strategy until one is found with a higher payoff than the current strategy. Hence, it does not require that the player know the utility of every strategy in her strategy space.

As with the best reply dynamic, it is fairly straightforward to show that if the random better reply dynamic converges, then it must converge to a Nash Equilibrium. It is more difficult to show that the random better reply dynamic will guarantee that the play will move into $\mathbf{D}^\infty(\mathbf{S})$ unless the game is finite.

Two more definitions will be useful to us in this chapter.

Definition 23. *A sequence of strategy profiles, $\{\mathbf{s}^0, \mathbf{s}^1, \mathbf{s}^2, \ldots\}$, is called a path if for each $k > 0$ there exists a player i_k such that $\mathbf{s}_{-i_k}^{k-1} = \mathbf{s}_{-i_k}^k$. In other words, at each stage at most one player can change strategies.*

Definition 24. *A path $\{\mathbf{s}^0, \mathbf{s}^1, \mathbf{s}^2, \ldots\}$ is called an improvement path if for each $k > 0$ (and i_k as defined above), $u_{i_k}(\mathbf{s}^k) > u_{i_k}(\mathbf{s}^{k-1})$. In other words, a path is an improvement path if the player that changes strategies always improves her utility.*

Note that both of the dynamics we have defined, the best reply and better reply dynamics, will define paths through the space of strategy profiles. Furthermore, if we eliminate steps at which players do not change strategies, the paths defined will be improvement paths. (If the dynamic converges in a finite number of steps, then eliminating such steps will produce a finite path.)

We now define a special class of games: potential games.

5.2 POTENTIAL GAMES

Potential games were defined and their properties discussed in [48]. If a game is constructed with random payoffs, then the probability that it will be a potential game is zero. Hence, in some sense, potential games are rare. Nevertheless, potential games seem to appear with surprising regularity when real-world situations are modeled. We will see an example of such a game at the end of this chapter.

5.2.1 Definition and Basic Properties

We begin, of course, by defining what we mean by a potential game.

Definition 25. *A game* $\Gamma = (\mathbf{N}, \mathbf{S}, \{u_i\})$ *is an* exact potential game *if there exists a function* $V : \mathbf{S} \to \mathbb{R}$ *such that for all* $i \in \mathbf{N}$, *all* $s \in \mathbf{S}$, *and all* $s_i' \in \mathbf{S_i}$,

$$V(s_i, \mathbf{s_{-i}}) - V(s_i', \mathbf{s_{-i}}) = u_i(s_i, \mathbf{s_{-i}}) - u_i(s_i', \mathbf{s_{-i}}).$$

The function V *is called an exact potential function for the game* Γ.

Definition 26. *A game* $\Gamma = (\mathbf{N}, \mathbf{S}, \{u_i\})$ *is an* ordinal potential game *if there exists a function* $V : \mathbf{S} \to \mathbb{R}$ *such that for all* $i \in \mathbf{N}$, *all* $\mathbf{s} \in \mathbf{S}$, *and all* $s_i' \in \mathbf{S_i}$,

$$V(s_i, \mathbf{s_{-i}}) - V(s_i', \mathbf{s_{-i}}) > 0 \Leftrightarrow u_i(s_i, \mathbf{s_{-i}}) - u_i(s_i', \mathbf{s_{-i}}) > 0.$$

The function V *is called an* ordinal potential function *for the game* Γ.

In both cases, note that the potential function reflects the change in utility for any unilaterally deviating player. Since unilateral deviations are central to the Nash equilibrium, the following result should come as no surprise. Also note that every exact potential game is obviously an ordinal potential game with the same potential function V.

Theorem 15. *If* V *is an ordinal potential function of the game* Γ *and* $\mathbf{s}^* \in \{\arg \max_{\mathbf{s}' \in \mathbf{S}} V(\mathbf{s}')\}$ *is a maximizer of the potential function, then* $\mathbf{s}^*$ *is a Nash Equilibrium of the game.*

The following corollary is immediate.

Theorem 16. *If* Γ *is a finite ordinal potential game, then* Γ *has at least one Nash Equilibrium in pure strategies.*

If the game is infinite, then we require slightly more. The following corollary is an immediate consequence of the well-known Weierstrauss theorem in optimization theory.

Theorem 17. *If Γ is an ordinal potential game with a compact strategy space $\mathbf{S}$ and a continuous potential function V, then Γ has at least one Nash Equilibrium in pure strategies.*

5.2.2 Convergence

Now, we can develop the most important results on potential games. We start by defining a property.

Definition 27. *A game Γ has the* finite improvement path property *if every improvement path in Γ is of finite length.*

Theorem 18. *All finite ordinal potential games have the finite improvement path property.*

Proof. Let $\{\mathbf{s}^0, \mathbf{s}^1, \mathbf{s}^2, \ldots\}$ be an improvement path. Then, since for all $k > 0$, $u_{i_k}(\mathbf{s}^k) > u_{i_k}(\mathbf{s}^{k-1})$, it follows that $V(\mathbf{s}^k) > V(\mathbf{s}^{k-1})$ for all $k > 0$. So $\{V(\mathbf{s}^0), V(\mathbf{s}^1), V(\mathbf{s}^2), \ldots\}$ is a strictly increasing sequence. But $\mathbf{S}$ is a finite set, hence the sequence $\{\mathbf{s}^0, \mathbf{s}^1, \mathbf{s}^2, \ldots\}$ must be finite. $\square$

But, if every improvement path is finite and the best and random better response algorithms trace out improvement paths which continue until they converge (at which point we must be at a Nash equilibrium), then we can immediately state the following theorem.

Theorem 19. *Let Γ be a finite ordinal potential game. Then both the best reply dynamic and the random better reply dynamic will (almost surely) converge to a Nash Equilibrium in a finite number of steps.*

The stipulation that the convergence is "almost sure" comes from our allowance that nodes might be chosen at random gives rise to the probability zero event that we will choose the same node over and over again until the end of time. This might cause the algorithm never to converge to a Nash equilibrium, but the probability that this will happen is zero.

The good news, though, is that the convergence of these algorithms does not seem to require that the games be finite. (Proving convergence in this case requires a bit more work, however.) We begin with Zangwill's convergence theorem [49].

Theorem 20 (Zangwill). *Let $\mathbf{f} : \mathbf{X} \rightrightarrows \mathbf{X}$ determine an algorithm that, given a point x^0, generates a sequence $\{x^k\}$ through the iteration $x^{k+1} \in \mathbf{f}(x^k)$. Let a solution set, $\mathbf{S}^* \subset \mathbf{X}$ be given. Suppose*

1. *all points $\{x^k\}$ are in a compact set $\mathbf{S} \subset \mathbf{X}$;*
2. *there is a continuous function $\alpha : \mathbf{X} \to \mathbb{R}$ such that:*
    (a) if $x \notin \mathbf{S}^*$, then $\alpha(x') > \alpha(x)$, $\forall x' \in \mathbf{f}(x)$;
    (b) if $x \in \mathbf{S}^*$, then $\alpha(x') \geq \alpha(x)$, $\forall x' \in \mathbf{f}(x)$; and
3. *$\mathbf{f}$ is closed at x if $x' \notin \mathbf{S}^*$.*

Then either $\{x^k\}$ arrives at a solution in $\mathbf{S}^*$ or every limit point of $\{x^k\}$ is in $\mathbf{S}^*$.

Theorem 21. *Let Γ be an ordinal potential game with a compact action space $\mathbf{S}$ and a continuous potential function V. Then the best response dynamic will (almost surely) either converge to a Nash Equilibrium or every limit point of the sequence will be a Nash Equilibrium.*

Proof.   Given the compact strategy space $\mathbf{S}$, let $\mathbf{f(s)}$ be the set of all strategies in which exactly one player changes strategies to a best response. Then, if we take $\alpha = V$ and $\mathbf{S}^*$ to be the set of Nash Equilibria, the best response algorithm generates a sequence of strategy profiles satisfying the requirements of Zangwell's convergence theorem, provided that we ignore steps at which the chosen user does not change strategies.

Ignoring the steps at which the chosen user does not change strategies will only create a problem if their are an infinite number of such steps in a row. This occurs only with probability zero unless we are at a Nash Equilibrium.   □

A similar result is believed to be true for the random better response dynamic, but the authors are not aware of a formal proof.

5.2.3   Identification

If one is provided with a potential function, it is easy to check whether or not a given game is a potential game. If you have only a game model and no potential function, how can you recognize a potential game? For ordinal potential games, we can provide no assistance. For exact potential games, though, we provide some properties that might be helpful in recognizing these games.

Definition 28. *A coordination game is a game in which all users have the same utility function. That is, $u_i(\mathbf{s}) = C(\mathbf{s})$ for all $i \in \mathbf{N}$.*

Note that a coordination game is immediately an exact potential game with potential function $V = C$.

Definition 29. *A dummy game is a game in which each player's payoff is a function of only the actions of other players (i.e., her own action does not affect her payoff). That is, for each i, $u_i(\mathbf{a}) = D_i(\mathbf{a_{-i}})$.*

Note that dummy games are also exact potential games where we take $V(\mathbf{s}) = 0$ (or any other constant function). Now, the somewhat surprising result is that any exact potential game can be written as the sum of a coordination game and a dummy game.

Theorem 22. *The game Γ is an exact potential game if and only if there exist functions $C : \mathbf{S} \to \mathbb{R}$ and $D_i : \mathbf{S_{-i}} \to \mathbb{R}$ such that $u_i(\mathbf{s}) = C(\mathbf{s}) + D_i(\mathbf{s_{-i}})$ for all i and all $\mathbf{s}$. The exact potential function of this game is $V(\mathbf{s}) = C(\mathbf{s})$.*

So, one way of identifying an exact potential game is to seek out this structure by trying to separate the game into a coordination game and a dummy game.

If the action spaces are intervals of the real line and the utility functions are twice continuously differentiable, then [48] gives a very straightforward way of identifying exact potential games.

Theorem 23.   *Let* Γ *be a game in which the strategy sets are intervals of real numbers. Suppose the payoff functions are twice continuously differentiable. Then* Γ *is a potential game if and only if*

$$\frac{\partial^2 u_i}{\partial s_i \partial s_j} = \frac{\partial^2 u_j}{\partial s_i \partial s_j}$$

for all i and j.

5.2.4   Interpretation

Before moving on to specific applications, we wish to provide a bit of interpretation. When modeling a communications or networking problem as a game, it is usually reasonably easy to determine the set of players and the strategy space, $\mathbf{S}$. The more difficult challenge is identifying the appropriate utility functions, u_i. Nevertheless, at the very least one can usually ascribe certain properties to the utility function, for example, whether it is decreasing or increasing in certain variables, whether it is convex or concave, etc.

In engineering applications, it is often also possible to identify an appropriate *social welfare function*, $W: \mathbf{S} \to \mathbb{R}$, which reflects the preferences of the system designer. For instance, the system designer might wish to maximize the total capacity of the wireless system or to minimize the probability that any individual node experiences a service outage.

If the social welfare function is an ordinal or exact potential function for the game, then game theory can provide strong results on the convergence of simple dynamics to desirable (social welfare maximizing) outcomes. More typically, though, there may be conflict between the socially desirable outcomes and the equilibria of the game. In these cases, we can either accept the suboptimal outcomes (if, for example, achieving an optimal outcome would be too costly) or modify our game (for instance, by moving to a repeated game framework where players are not myopic and introducing explicit or implicit reputation mechanisms) to try to induce better outcomes.

5.3   APPLICATION: INTERFERENCE AVOIDANCE

In a code division multiple access (CDMA) system, users are differentiated by their choice of different spreading codes. Each user's spreading code can be viewed as a strategy. The payoff of this strategy is related to its orthogonality with (or, rather, lack of correlation with) the spreading codes employed by other users. As with all potential games, note that this game has an aspect of "common good." Whenever a player changes her strategy in a way that improves her payoff (by making her own code less correlated with the other codes in the system), this change will also improve the payoffs of the other players (as their strategies will now be less correlated with hers).

To briefly formalize this notion in a relatively simple setting, suppose that each player chooses a code from $\mathbf{S_i} = \{\mathbf{s_i} \in \mathbb{R}^M | \|\mathbf{s_i}\| = 1\}$ where $\|\mathbf{s_i}\|$ represents the usual vector norm in $\mathbb{R}^M$ and M is the spreading factor of the system. (The constraint that $\|\mathbf{s_i}\| = 1$ is a constraint on nodes' transmit powers. For this simple example, we assume that all transmissions have the same power at the receiver.) Assume that $\mathbf{s_i}$ is written as a column vector. And rather than our usual definition for $\mathbf{S}$ as the strategy space, for the purpose of this example, let $\mathbf{S}$ be an $M \times |\mathbf{N}|$ matrix whose ith column is $\mathbf{s_i}$. Let $\mathbf{S_{-i}}$ represent this matrix with the ith column removed. Also part of the system model is additive M-dimensional white Gaussian noise, assumed to have an $M \times M$ covariance matrix $\mathbf{R_z}$.

We can then define a number of different utility functions, the exact form of which depends on the metric of success (signal-to-interference-and-noise ratio or mean squared error) and the type of receiver (correlation receiver or maximum signal to interference and noise ratio receiver). One such utility function would be the signal to interference and noise ratio (SINR) for a correlation receiver, which is given by:

$$u_i(\mathbf{S}) = \frac{1}{\mathbf{s_i}^T R_i \mathbf{s_i}}.$$

In this expression, $R_i = \mathbf{S_{-i}}\mathbf{S_{-i}}^T + \mathbf{R_z}$, which is the autocorrelation matrix of the interference plus noise.

It then turns out that the so-called negated generalized total squared correlation function, which is defined as $V(\mathbf{S}) = -\|\mathbf{SS}^T + \mathbf{R_z}\|_F^2$, is an ordinal potential function for this game (and several other interference avoidance games). (The notation $\|\cdot\|_F$ represents the Frobenius matrix norm, which is the square root of the sum of the squared matrix entries.)

The somewhat surprising result of this formulation of the interference avoidance game is that allowing nodes to greedily choose the best spreading codes for themselves leads to a set of spreading codes that minimizes generalized total squared correlation (which in some cases represents a set of codes that maximizes the sum capacity of the system). For more information about this example, see [50]. For more information about interference avoidance in general, see [51].

CHAPTER 6

Future Directions

The previous five chapters have outlined fundamental concepts that the researcher in wireless communications and networks is likely to encounter as she starts exploring the field of game theory. We also discussed some examples of game theoretic models that were developed to better understand issues in power control, medium access control, topology formation, node participation, interference avoidance, etc.

This treatment is by no means comprehensive, as game theory is a vast field and networking and communications researchers have applied it to a large number of problems. Among those we did not discuss in detail are routing, quality of service (QoS), and TCP congestion control. A more complete survey of applications of game theory to wireless systems, and in particular wireless ad hoc networks, can be found in [52].

In this chapter, we focus on emerging research issues on wireless communications and wireless networks where we feel game theory still has an important role to play.

6.1 RELATED RESEARCH ON WIRELESS COMMUNICATIONS AND NETWORKING

Let us now discuss some emerging research issues and our vision of how game theory may help further our understanding of those issues.

6.1.1 The Role of Information on Distributed Decisions

As wireless networks evolve, we are seeing a tendency toward decentralized networks, in which each node may play multiple roles at different times without relying on a base station or access point to make decisions such as how much power to use during transmission, in what frequency band to operate, etc. Examples include sensor networks, mobile ad hoc networks, and networks used for pervasive computing. These networks are also self-organizing and support multihop communications.

These characteristics lead to the need for distributed decision making that potentially takes into account network conditions as well as channel conditions. An individual node may need to have access to control information regarding other nodes' actions, network congestion, etc.

How much information is enough for effective distributed decision making? The performance difference between following selfish, local goals and acting in a communal, network-wide mode of operation is called the *price of anarchy* in [53]. Game theory may also help us assess this cost.

In most of this book, we assumed that all players knew others' utility functions and, in the case of repeated games, others' actions in previous rounds. This is clearly not the case in many networking problems of interest. Several game theory formulations are possible to account for this uncertainty.

Think back about our discussion of games in extensive form in Chapter 4. A game is said to be of perfect information if each information set is a singleton; otherwise, the game is a *game of imperfect information*. This can be used to model ignorance about other nodes' types (for instance, other nodes in a wireless network may have very different utility functions) or other nodes' actions. In games of imperfect information, the Nash equilibrium definition is refined to reflect uncertainties about players' types, leading to the concept of Bayesian equilibrium. This type of formulation has been applied, for example, to the study of disruptions caused by a rogue node in an ad hoc network [27].

Another model formulation that accounts for uncertainties regarding other players' actions is that of *games of imperfect monitoring*. In a repeated game with perfect monitoring, each player directly observes the actions of every other player at the end of each stage. However, such an assumption of perfect monitoring is challenged in situations where an opponent's actions cannot be directly monitored. For example, in a node participation game for ad hoc networks, nodes could deny forwarding packets for others (or only occasionally forward packets) to conserve their limited resources. It is infeasible in these situations for a node to directly monitor the actions of its neighbors.

To account for the lack of perfect monitoring, we can study repeated games with imperfect public monitoring [54]. Here a player, instead of monitoring her opponents' actions, observes a random public signal at the end of every stage of the game. This public signal is correlated to the actions of all other players in the game but it does not reveal deterministically what these actions were. Specifically, the distribution function of the public signal depends on the action profile chosen by the other players. Typically, it is assumed that the support of the public signal distribution is constant across all possible joint actions. In the node participation example, a possible signal could be the network goodput as perceived by each node in the network. The analysis of an imperfect monitoring game involves finding a perfect public equilibrium strategy that serves as the solution of the game.

Recently, game theoreticians have also studied games with private monitoring. In such games, each player has a distinct, individual assessment of others' likely actions in previous stages. The analysis of outcomes (and of issues such as collusion) for those types of games is still in its infancy.

6.1.2   Cognitive Radios and Learning

Cognitive radios [55] are currently viewed as a promising approach to make better use of the wireless spectrum by rapidly adapting to changes in the wireless channel. We envision this concept to eventually extend to cognitive networks, where each node individually adapts based on network-wide considerations to yield better overall performance to the network as a whole [56].

In this context, network nodes are expected to be aware of their environment, to be able to dynamically adapt to that environment, and to be able to learn from outcomes of past decisions. It can be argued that, to a varying extent, today's radios already perform the first two functions, as even something as simple as a cordless phone is able to detect channel use and select the appropriate channel accordingly. Both more sophisticated adaptation and the ability to learn are expected from future radios, to solve complex problems such as the opportunistic utilization of spectrum.

The study of how these radios will learn and how fast they will converge to effective solutions becomes relevant. The Artificial Intelligence community already employs game theory in understanding strategic and cooperative interactions in multiagent systems. Very recently, there has been work on applying game theory to the analysis of the performance of networks of cognitive radios [57], but there is still much to be done in this area.

6.1.3   Emergent Behavior

Emergent behavior can be defined as behavior that is exhibited by a group of individuals that cannot be ascribed to any particular member of the group. Flocking by a group of birds or anthill building by a colony of ants is common example. Emergent behavior has no central control and yet often results in benefits to the group as a whole: for instance, cities self-organize into complex systems of roads, neighborhoods, commercial centers, etc., to the benefit of their inhabitants.

The observation of emergent behavior in ants and other social insects indicates that simple decisions lead to global efficiency (emergent systems get unwieldy if individual behavior is too complex), and that local information can lead to global wisdom [58].

Translating into the realm of wireless networks, it is important to understand whether simple strategies by individual nodes will lead to globally optimal solutions and whether nodes can learn enough from their neighbors to make effective decisions about the network as a whole.

As it is very difficult to experiment with large decentralized wireless networks (containing hundreds or thousands of nodes), an analytical approach to understand emergent behavior in such networks is highly desirable. Game theory is one promising approach toward such an understanding.

6.1.4   Mechanism Design

Mechanism design is an area of game theory that concerns itself with how to engineer incentive mechanisms that will lead independent, self-interested participants toward outcomes that are desirable from a systemwide point of view.

As we learn through game theory that selfish behavior by users may lead to inefficient equilibria (recall once again some of the examples of single-stage games), it becomes important to build incentive structures that will induce more desirable equilibria. Such incentives can be explicit, using for example pricing or virtual currency mechanisms, or implicit in the choice of utility functions.

One important distinction between the application of game theory to communications and networks and its more traditional application to economic problems is that in the latter strict rationality assumptions may not hold. On the other hand, in a network we often have the opportunity to program nodes with some objective function to maximize. As these nodes proceed to make decisions without human intervention, they can be assumed to be perfectly rational.

In more traditional economic and social problems, it is difficult to predict the utility function that players are following (such a utility function may not be clearly known to players themselves). In networking problems, we may be able to program nodes with a utility function. The problem, however, becomes identifying the appropriate utility function that will lead to a desirable equilibrium.

6.1.5   Modeling of Mobility

Few existing game theoretic models of wireless networks consider the effects of mobility on players' decisions. In some cases, it is reasonable to abstract mobility considerations. For instance, a single stage game can represent a snapshot of the current network, so mobility need not be considered. Even in a repeated game, one can implicitly recognize that there is mobility by assigning a probability that each node will participate in the current round of the game.

However, some considerations, such as convergence to an efficient equilibrium or learning in a network, do require that the model be tied to a realistic mobility model. After all, it may be difficult to learn from others' actions, or to devise credible threats, if the set of nodes with whom a player interacts significantly changes from stage to stage of the game. A comprehensive characterization of the effects of mobility on the outcomes of the game is an open area of research.

6.1.6   Cooperation in Wireless Systems and Networks

There is increasing interest in cooperation in wireless systems and networks, at all layers of the protocol stack. This includes cooperative communications, exchange of trust and reputation information, forwarding in ad hoc networks, etc. Cooperative game theory can help us better understand some of the issues involved in cooperation and bargaining.

One example is the issue of dynamic access to the available spectrum. It is well known that the wireless spectrum is inefficiently utilized; one solution to mitigate this problem would be to allow prospective users of the spectrum (including, but not limited to, current license holders) to negotiate its use dynamically, based on current traffic requirements and channel availability. It is possible to model this scenario as a cooperative game, and to develop distributed bargaining mechanisms that are expected to lead to a fair and stable outcome.

6.2 CONCLUSIONS

Game theory is a fascinating field of study. In this work, we have barely scratched the surface. We hope that the reader will now have an understanding of the fundamental concepts in game theory that will enable him or her to pursue a more in-depth study of the available literature. At times, we have simplified some of the concepts and results rather than present them in all possible generality. We suspect a seasoned game theoretician may be bothered by some of those simplifications, but we hope the engineer who is a novice to the field of game theory will forgive the omissions: they were made in the interest of clarity.

Despite the recent popularity of game theory in communications and networking research, the potential for further findings is vast. We outline above some areas of study that we believe are wide open for further exploration.

Most important, engineers must be aware of the limitations and assumptions behind game theoretic models. We do not argue that game theory is the appropriate tool to understand all, or even most, communications and networking phenomena. We do believe that it is a promising tool to understand an important class of network issues, particularly those involving interdependent adaptations.

Bibliography

[1] D. Fudenberg and J Tirole, *Game Theory*. Cambridge, MA: MIT Press, 1991.

[2] M. J. Osborne and A. Rubinstein, *A Course in Game Theory*. Cambridge, MA: MIT Press, 1994.

[3] R. J. Aumann and M. Maschler, "Game theoretic analysis of a bankruptcy problem from the talmud," *J. Econ. Theory*, vol. 36, pp. 195–213, 1985. doi:10.1016/0022-0531(85)90102-4

[4] P. Walker, "An outline of the history of game theory." Available at: http://william-king.www.drekel.edu/top/class/histf.html, April 1995.

[5] J. F. Nash, "Equilibrium points in n-person games," *Proc. Natl. Acad. Sci. U.S.A.*, vol. 36, no. 1, pp. 48–49, January 1950. [Online]. Available at: http://links.jstor.org/sici?sici=0027-8424%2819500115%2936%3A1%3C48%3AEPING%3E2.0.CO%3B2-B.

[6] J. John F. Nash, "The bargaining problem," *Econometrica*, vol. 18, no. 2, pp. 155–162, April 1950. [Online]. Available: http://links.jstor.org/sici?sici=0012-9682%2819504%2918%3A2%3C155%3ATBP%3E2.0CO%3B2-H

[7] J. Nash, "Two-person cooperative games," *Econometrica*, vol. 21, no. 1, pp. 128–140, January 1953. [Online]. Available at: http://links.jstor.org/sici?sici=0012-9682%2819530101%2921% 3A1%3C128%3ATCG%3E2.0.CO%3B2-N

[8] "About RAND: History and mission." Available at: http://www.rand.org/about/history/.

[9] H. Ji and C. -Y. Huang, "Non-cooperative uplink power control in cellular radio systems," *Wireless Networks*, vol. 4, no. 3, pp. 233–240, 1998. doi:10.1023/A:1019108223561

[10] D. Goodman and N. Mandayam, "Power control for wireless data," *IEEE Pers. Communications Magazine*, vol. 7, no. 2, pp. 48–54, April 2000.doi:10.1109/98.839331

[11] C. U. Saraydar, N. B. Mandayam, and D. J. Goodman, "Efficient power control via pricing in wireless data networks," *IEEE Trans. Communications*, vol. 50, no. 2, pp. 291–303, February 2002.doi:10.1109/26.983324

[12] T. Heikkinen, "Distributed scheduling via pricing in a communication network," in *Proceedings of Networking*. Springer-Verlag, May 2002.

[13] T. Alpcan, T. Basar, R. Srikant, and E. Altman, "CDMA uplink power control as a noncooperative game," in *Proc. IEEE Conference on Decision and Control*, Orlando, FL, 2001, pp. 197–202.

[14] M. Xiao, N. Schroff, and E. Chong, "Utility based power control in cellular radio systems," in *Proceedings of IEEE INFOCOM*, Anchorage, Alaska, 2001.

[15] A. Economides and J. Silvester, "Multi-objective routing in integrated services networks: A game theory approach," in *IEEE INFOCOM Networking in the 90s/Proceedings of the 10th Annual Joint Conference of the IEEE and Communications Societies*, vol. 3, 1991, pp. 1220–1227.

[16] Y. Korilis, A. Lazar, and A. Orda, "Achieving network optima using Stackelberg routing strategies," *IEEE/ACM Trans. Networking*, vol. 5, no. 1, pp. 161–173, February 1997.doi:10.1109/90.554730

[17] R. J. La and V. Anantharam, "Optimal routing control: Game theoretic approach," in *Proceedings of the 36th IEEE Conference on Decision and Control*, vol. 3, 1997, pp. 2910–2915.doi:full_text

[18] T. Roughgarden and E. Tardos, "How bad is selfish routing?" *Journal of the ACM*, vol. 49, no. 2, pp. 236–259, March 2002.doi:10.1145/506147.506153

[19] D. Braess, "Uber ein paradoxen der verkehrsplanung," *Unternehmenforschung*, vol. 12, pp. 258–268, 1968.

[20] J. D. Murchland, "Braess's paradox of traffic flow," *Transport. Res.*, vol. 4, pp. 391–394, 1970.doi:10.1016/0041-1647(70)90196-6

[21] P. Michiardi and R. Molva, "Core: A collaborative reputation mechanism to enforce node cooperation in mobile ad hoc networks," in *Proc. of the 6th Joint Working Conference on Communications and Multimedia Security*, September 2002, pp. 107–121.

[22] S. Buchegger and J. Y. LeBoudec, "Performance analysis of the CONFIDANT protocol: Cooperation of nodes—fairness in dynamic ad hoc networks," in *Proceedings of the ACM MobiHoc*, June 2002.

[23] P. Dewan and P. Dasgupta, "Trusting routers and relays in ad hoc networks," in *Proceedings of the International Conference on Parallel Processing Workshops*, October 2003, pp. 351–359.

[24] P. Dewan, P. Dasgupta, and A. Bhattacharya, "On using reputations in ad hoc networks to counter malicious nodes," in *Proceedings of the 10th International Conference on Parallel and Distributed Systems*, July 2004, pp. 665–672.

[25] J. Liu and V. Issarny, "Enhanced reputation mechanism for mobile ad hoc networks," in *Proceedings of the 2nd International Conference on Trust Management*, April 2004.

[26] M. T. Refaei, V. Srivastava, L. A. DaSilva, and M. Eltoweissy, "A reputation-based mechanism for isolating selfish nodes in ad hoc networks," in *Proceedings of the 2nd Annual International Conference on Mobile and Ubiquitous Systems (MobiQuitous): Networking and Services*, San Diego, CA, July 2005.

[27] L. A. DaSilva and V. Srivastava, "Node participation in ad-hoc and peer-to-peer

networks: A game-theoretic formulation," in *Workshop on Games and Emergent Behavior in Distributed Computing Environments*, Birmingham, U K, September 2004.

[28] P. Michiardi and R. Molva, "Game theoretic analysis of security in mobile ad hoc networks," Institut Eurécom, Technical Report RR-02-070, April 2002.

[29] P. Michiardi and R. Molva, "Analysis of coalition formation and cooperation strategies in mobile ad hoc networks," *Ad Hoc Networks*, vol. 3, no. 2, pp. 193–219, March 2005.doi:10.1016/j.adhoc.2004.07.006

[30] M. Allais, "Le comportement de l'homme rationnel devant de risque: Critique des postulats et axiomes de l'ecole americaine," *Econometrica*, vol. 21 pp. 503–546, 1953.

[31] D. Kahneman and A. Tversky, "Prospect theory: An analysis of decision under risk," *Econometrica*, vol. 47, pp. 263–291, 1979.

[32] D. M. Kreps, *Notes on the Theory of Choice*, ser. Underground Classics in Economics. Westview Press, 1988.

[33] P. Fishburn, *Utility Theory for Decision Making*. New York: John Wiley and Sons, 1970.

[34] L. J. Savage, *The Foundations of Statistics*. New York: Wiley, 1954.

[35] F. J. Anscombe and R. J. Aumann, "A definition of subjective probability," *Ann. Math. Statist.*, vol. 34 pp. 199–205, 1963.

[36] D. Ellsberg, "Risk, ambiguity, and the savage axioms," *Q. J. Econ.*, vol. 75, no. 4, pp. 643–669, November 1961.

[37] W. Poundstone, *Prisoner's Dilemma: John von Neumann, Game Theory, and the Puzzle of the Bomb*. Doubleday, 1992.

[38] J. Ratliff, "Jim Ratliff's graduate-level course in game theory," available on Jim Ratliff's Web site, 1992–1997 [Online]. Available at: http://virtualperfection.com/gametheory/.

[39] L. A. DaSilva, "Static pricing in multiple-service networks: A game-theoretic approach," Ph. D. dissertation, the University of Kansas, 1998.

[40] R. Cocchi, S. J. Shenker, D. Estrin, and L. Zhang, "Pricing in computer networks: Motivation, formulation, and example," *IEEE/ACM Trans. Networking*, vol. 1 no. 6, pp. 614–627, 1998.

[41] A. A. Lazar, A. Orda, and D. E. Pendarakis, "Virtual path bandwidth allocation in multi-user networks," in *Proceedings of IEEE INFOCOM*, 1995, pp. 312–320.

[42] Y. Qiu and P. Marbach, "Bandwidth allocation in ad hoc networks: A price-based approach," in *Proceedings of IEEE INFOCOM*, vol. 2, 2003, pp. 797–807.

[43] Y. Xue, B. Li, and K. Nahrstedt, "Price based resource allocation in wireless ad hoc networks," in *Proceedings of 11th International Workshop on Quality of Service*, 2003.

[44] Y. A. Korilis and A. A. Lazar, "On the existence of equilibria in noncooperative optimal flow control," *J. ACM*, 1995.

[45] A. B. MacKenzie and S. B. Wicker, "Game theory in communications: Motivation, explanation, and application to power control," in *Proceedings of IEEE GLOBECOM*, 2001, pp. 821-826.

[46] V. Shah, N. B. Mandayam, and D. J. Goodman, "Power control for wireless data based on utility and pricing," in *Proceedings of the 9th IEEE International Symposium on Personal, Indoor and Mobile Radio Communications*, September 1998, pp. 1427–1432.

[47] A. B. MacKenzie and S. B. Wicker, "Stability of multipacket slotted aloha with selfish users and perfect information," in *Proceedings of IEEE INFOCOM*, April 2001.

[48] D. Monderer and L. S. Shapley, "Potential games," *Games Econ. Behav.*, vol. 14, no. 1, pp. 124–143, 1996.doi:10.1006/game.1996.0044

[49] W. Zangwill, *Nonlinear Programming: A Unified Approach*. Englewood Cliffs, NJ: Prentice-Hall, 1969.

[50] J. E. Hicks, A. B. MacKenzie, J. A. Neel, and J. H. Reed, "A game theory perspective on interference avoidance," in *Proceedings of the Global Telecommunications Conference (Globecom)*, vol. 1, 2004, pp. 257–261.doi:full_text

[51] D. C. Popescu and C. Rose, *Interference Avoidance Methods for Wireless Systems*. Kluwer Academic/Plenum Publishers, 2004.

[52] V. Srivastava, J. Neel, A. B. MacKenzie, R. Menon, L. A. DaSilva, J. E. Hicks, J. H. Reed, and R. P. Gilles, "Using game theory to analyze wireless ad hoc networks," *IEEE Communication Surveys and Tutorials*, 2005.

[53] C. H. Papadimitriou, "Algorithms, games and the internet," in *Proceedings of the 33rd ACM Symposium on Theory of Computing*, 2001, pp. 749–753.

[54] D. Abreu, D. Pearce, and E. Stachetti, "Toward a theory of discounted repeated games with imperfect monitoring," *Econometrica*, vol. 58, no. 6, pp. 1041–1063, September 1990.

[55] S. Haykin, "Cognitive radio: Brain-empowered wireless communications," *IEEE J. Selected Areas Communications*, vol. 23, no. 2, pp. 201–220, February, 2005. doi:10.1109/JSAC.2004.839380

[56] R. W. Thomas, L. A. DaSilva, and A. B. MacKenzie, "Cognitive networks," in *Proceedings of First IEEE International Symposium on New Frontiers in Dynamic Spectrum Access Networks (DySPAN 05)*, November 2005.

[57] J. Neel, R. M. Buehrer, J. H. Reed, and R. P. Gilles, "Game theoretic analysis of a network of cognitive radios," in *Proceedings of the 45th Midwest Symposium on Circuits and Systems*, vol. 3, August 2002, pp. 409–412.

[58] D. Gordon, *Ants at Work: How an Insect Society is Organized*. Free Press, 1999.

Author Biographies

Allen B. MacKenzie Allen B. MacKenzie received the B.Eng. degree from Vanderbilt University, Nashville, TN in 1999 and the Ph.D. degree from Cornell University, Ithaca, NY in 2003. Since 2003, he has been an Assistant Professor in the Bradley Department of Electrical and Computer Engineering at Virginia Polytechnic Institute and State University, Blacksburg, VA. His Ph.D. dissertation was on the game-theoretic modeling of power control and medium access control problems, and his current research interests are in the areas of cognitive radio, cognitive networks, and the analysis and design of wireless networks using techniques from game theory. He received a 2005 National Science Foundation CAREER award to continue the development of game theoretic models of wireless networks. He has other ongoing research projects in the areas of cognitive radio and ad hoc networks sponsored by the National Science Foundation and the National Institute of Justice. He serves on the technical program committee of several international conferences in the areas of communications and networking, and is a regular reviewer for journals in these areas. He is affiliated with both the Center for Wireless Telecommunications (CWT) and the Mobile and Portable RadioResearch Group (MPRG) at Virginia Tech. He is a member of the IEEE, a member of the ACM, and a member of the ASEE. He frequently teaches courses on communications theory and networking.

Luiz A. DaSilva Luiz A. DaSilva joined Virginia Tech's Bradley Department of Electrical and Computer Engineering in 1998, where he is now an Associate Professor. He received his Ph.D. in Electrical Engineering from the University of Kansas and previously worked for IBM for 6 years. His research interests focus on performance and resource management in wireless mobile networks and Quality of Service (QoS) issues. He is currently involved in funded research in the areas of cognitive networks, application of game theory to model mobile ad-hoc networks (MANETs), heterogeneous MANETs employing smart antennas, and distributed trust and reputation management, among others. He has published more than 50 refereed papers in journals and major conferences in the communications and computer areas. Current and recent research sponsors include NSF, the Office of Naval Research, Booz Allen Hamilton, the U.S. Customs Services, Intel, and Microsoft Research, among others. He serves in technical program committees for numerous conferences and was the general chair for the 14th International Conference on Computer Communications and Networks. He is a member of the Center for Wireless Communications (CWT), associated faculty at the Mobile and Portable Radio Research Group (MPRG), and a member of the Governing Board of the NSF-funded Integrated Research and Education in Advanced Networking (IREAN) program at Virginia Tech. He is a senior member of IEEE, a member of ASEE, and a past recipient of the ASEE/IEEE Frontiers in Education New Faculty Fellow award. He frequently teaches courses on network architecture and protocols and on mobile and wireless networking.

Printed in the United States
by Baker & Taylor Publisher Services